Elements of Agricultural Chemistry

By

R.K. Kaleeswari

R. Rajeswari

J. Prabhaharan

C. Bharathi

SSPH

SATISH SERIAL PUBLISHING HOUSE

403, Express Tower, Commercial Complex, Azadpur, Delhi-110033 (India)

Phone : 011-27672852, Fax : 91-11-27672046

E-mail : info@satishserial.com, hkjain1975@yahoo.com

Website : www.satishserial.com

Published by :

SATISH SERIAL PUBLISHING HOUSE

403, Express Tower, Commercial Complex, Azadpur, Delhi-110033 (INDIA)
Phone : 011-27672852 Fax : 91-11-27672046
E-mail : info@satishserial.com, hkjain1975@yahoo.com

ISBN 978-93-81226-87-2

Typeset at: Harminder Kharb *for* Laxmi Art Creation # 9971212252
Printed at: Salasar Imaging Systems, Delhi-110035

PREFACE

The rationale of this book is to offer a concise outline of the general principles of Agricultural chemistry. This anthology discusses the proximate and ultimate constituents of crops. The goals of agricultural chemistry are to expand the understanding of the causes and effects of biochemical reactions related to plant growth. Every scientific discipline that contributes to agricultural progress depends in some way on chemistry. Hence agricultural chemistry is not a distinct discipline, but a common cord that fasten together genetics, physiology, microbiology, entomology, and numerous other sciences that impinge on agriculture.

Biochemical pathways involved in the synthesis of starch, protein and fat are well portrayed for understanding the plant physiological activities. This book also covers the post harvest changes take place in crop produce for better handling in storage and transport. The chemical constituents of crops would be great assistance to the plant breeders for scheduling their research on plant nutrient basis.

Agrochemicals developed to lend a hand in the production of food, feed, and fiber include scores of herbicides, insecticides, fungicides and other pesticides, plant growth regulators, fertilizers, and animal feed supplements. Chief among these groups from the ecological point of view are synthetic pesticides. The knowledge on chemical composition, mode of action and efficient method of application to reduce the pesticide residue in soil or water is indispensable for a sustainable crop production. This book describes molecular and structural formulae of the plant protection chemicals and transformation of pesticides in soil - plant -water ecosystem that impart technical knowhow to the scientists and farmers for safety handling of pesticides.

Authors

CONTENTS

Chapter 1

Proximate and Ultimate Constituent of Plants

The living plant is composed of various organic compounds which in turn are made up of several inorganic compounds or elements. The organic compounds to which the plant can be resolved at first disintegration or disappearance are called as proximate constituents. The inorganic compounds to which the plants can be finally decompose or disintegrates are called as ultimate constituents.

Proximate constituents

Water	-	80 to 95%
Carbohydrates	-	
a. Starch and sugars	-	1 to 5%
b. Hemicellulose	-	10 to 30%
c. Cellulose	-	20 to 30%
Proteins	-	1 to 15%
Lipids	-	1 to 5%
Tannins, alkaloids and essential oils	-	1 to 2%
Plant growth substances(enzymes, vitamins and hormones)	-	Traces
Minerals	-	2 to 5%

Water

It is the most abundant ingredient of active plant cells and constitutes 85 to 95% of the fresh weight of actively growing tissues. The water content decreases with maturity.

Functions of water in plant system

1. It is an important raw material for the synthesis of organic compounds.

2. It acts as a medium in which all the biological reactions are carried out.
3. It gives biological structure to the plant system.
4. The nutrients are absorbed by the plants only in the form of dilute solutions.
5. Water is essential for the transport of nutrients or water soluble compounds from one part to another part of the plant system.
6. It gives turgor to the plant system.

Carbohydrates

It is one of the major components present in the plant system. The sugars and starches those are present in the plant system serves as the food material for the growing plants. The hemicellulose and cellulose constitute the major part of the plant cells. Among all the constituents carbohydrate occupies more than 60% of the plant on dry weight basis.

Proteins

Proteins are nothing but the organic nitrogenous compounds or substances that are present in the plant system. Next to carbohydrate protein occupies the major part in the plant system. Proteins are structural compounds occurring in the protoplasm and nucleus of the cell system.

Lipids

These compounds are only soluble in organic solvents. These are storage compounds occurring mainly in the seeds.

Plant pigments

The plant pigments are the compounds which give colour to the leaves, stems and flowers. Chlorophyll, carotein and xanthophylls are some of the common plant pigments that are occurring in the plant system.

Tannins

These are the astringent (binding) compound of the complex substances that are present in the barks of many plant species.

Alkaloid

These are the nitrogenous waste materials that occur during N metabolism in plants.

Essential oils

These are the flavour producing substances that are present in plants.

Plant growth substances

Enzymes

These are the organic biocatalyst which will catalyze the biochemical reactions in the plant.

Vitamins

These are the substances which are closely connected to the plant metabolism.

Hormones

These are the substances which will promote growth, fruit setting and many other useful activities.

Minerals

These are inorganic elements that are present in the plant system. It may be any element (P, Ca, Mg, K, Na, etc.)

H_2O	- 75%
Oxygen	- 10%
Carbon	- 11%
Hydrogen	- 2%
Ash	- 2%

Ultimate constituents

The ultimate constituents are nothing but inorganic elements which will make up the plant body like C, K, N, P, Ca, Mg, Na, Si and other trace elements.

Carbon

It is the basic element forming the skeleton of all organic compounds occurring in the plant system.

Hydrogen and Nitrogen

Along with C, the H and N constitutes more than 19% of the plant system. Most of the plant organic compounds contain C and N.

Silicon and sodium

These two elements will accumulate in the plant system in appreciable amount due to incidental absorption by the plant itself.

❑❑❑

Chapter 2

Nutritional Quality of Cereals

Rice

The nutritional value of the rice kernel in its natural state is comparable to the nutritive value of other cereals. The lemma and palea along with other associated structures viz., sterile lemma, rachilla and the awn together constitute the hull or husk. The hull constitutes approximately 20 to 25 per cent of the total weight of the rice. The average composition of the husked and milled rice is given below.

S. No.	Constituents	Husked	Milled
1.	Carbohydrates (%)	77.2	79.4
2.	Fat (%)	2.0	0.3
3.	Proteins (%)	8.9	7.6
4.	Ash (%)	1.9	0.4
5.	Calcium (%)	0.084	0.009
6.	Phosphorus (%)	0.290	0.096
7.	Iron	0.002	0.001

Vitamins (ppm)

1.	Thiamine (B_1)	3 to 5	0.6 to 1.0
2.	Riboflavin (B_2)	0.8 to 1.0	0.28
3.	Nicotinic acid	55	15 to 20
4.	Pantothenic acid	17	6.4

In glutinous rice varieties, the starch fraction is completely composed of ***'amylopectin'*** on the contrary in non-glutinous varieties it is composed of amylase and amylopectin. The sticky nature of the cooked rice is determined by the relative proportion of amylopectin and amylase. The main protein in rice is ***'oryzenin'***. Among the amino acids, the glutamic acid content is the highest. This is followed by aspartic acid and arginine. Rice grain is relatively high in lysine which constitutes about 4 per cent of the protein fraction. The nutritive value of rice is influenced by milling, storage conditions and genetic factors.

Wheat

It has the following composition

Starch	-	60 to 68 per cent
Protein	-	8 to 15 per cent
Fat	-	1.5 to 2 per cent
Cellulose	-	2 to 2.5 per cent
Minerals	-	1.5 to 2.0 per cent

It also contains vitamins A, B, B_2, nicotinic acid and vitamin E. The essential amino acids viz., isoleucine, leucine, lysine, methionine, phenyl alanin and valine are contained in the range of 26 to 30 per cent of the total amino acids. The main proteins present in the endosperm are gliaidine and glutenin, both of which are present in almost equal amounts. When water is added, they form a colloidal complex called gluten. The gluten in the bread wheat forms the typical spongy texture of bread.

Maize

On an average, maize grain contains

Carbohydrates	-	76 to 80 per cent
Protein	-	8 to 10 per cent
Fat	-	5 to 7 per cent
Fibre	-	3 to 5 per cent
Ash	-	2 per cent
Oil	-	4 to 4.5 per cent

The major portion of starch is amylopectin and the remainder in amylase. However, waxy maize wholly contains starch in the form of amylopectin. In the embryo protein consists of lysine and tryptophan and it is biologically balanced. However, protein in the endosperm does not have these essential amino acids. Recently released hybrids and composites have comparatively higher lysine content and the amino acids are better balanced. Low content of vitamin B complex, niacin and tryptophan in maize leads to a disease called ***"pellagra"*** in persons wholly dependent on maize. Vitamin A activity in yellow maize is primarily due to the colouring substances called *"cryptoxanthin"*.

Both the palatability and the nutritional quality of maize fodder are very high. It contains approximately 6 to 7 per cent crude protein and 85 to 95 per cent crude fibre. A high amount of free nitrates are also formed in maize plants. If the nitrate content exceeds 0.3 per cent then it may be fatal to animals.

Sorghum

The average composition of the sorghum grain is as follows:

Protein	-	7 to 14 per cent
Crude fat	-	2 to 7 per cent
Carbohydrates	-	70 to 90 per cent
Crude fibre	-	1 to 4 per cent
Calcium (mg/100g)	-	10 to 590 per cent
Phosphorus (mg/100g)	-	170 to 750 per cent
Iron (mg/100g)	-	1 to 20 per cent

In general, the self-pollinated varieties contain higher protein content than that of hybrids. Prolamine and glutamine are the main amino acids. Sorghum protein is deficient in arginine, lysine, glycine, tyrosine and methionine. Compared to the other cereals, sorghum protein is deficient in lysine and rich in leucine. Lysine is required for bone formation. Further, the tannins (polyphenol) in sorghum are associated with the poor utilization of the protein. Polyphenol levels in sorghum are higher in pigmented grains than those which are unpigmented.

Pearl millet

The mean composition of pearl millet grain is as follows:-

Carbohydrates	-	69.4 per cent
Protein	-	12 per cent
Crude fat	-	5 per cent
Fibre	-	2.4 per cent
Ash	-	2.3 per cent

The germ of pearl millet comprises about 17 per cent of the total kernel mass and contains about 25 per cent lipid, 20 per cent protein and most of the phytin, vitamins and enzymes. The protein content of the germ is of high quality containing higher levels of lysine, tryptophan and threonine. But in general, pearl millet has a low content of lysine, threonine and the sulphur containing amino acids. Pearl millet proteins have a higher digestibility than those of sorghum. The high oil content and enzyme activities of the germ have a major effect on storage properties and nutritional value of milled products. Flour storage ability is improved by the removal of germ but, nutritional value is reduced. The apparent amylase content of pearl millet starches ranges from 17 to 29 per cent.

Finger millet

The grain composition is as follows:-

Carbohydrate	-	72 per cent
Protein	-	7.3 per cent
Crude fat	-	1.3 per cent
Crude fibre	-	3.6 per cent
Minerals	-	2.7 per cent

It is nutritionally rich as it contains 2.7 per cent of minerals. It is the richest source of calcium and is also fairly rich in phosphorus as well as iron. The tannin content is the most significant antinutritional factor. But, it can be removed by hulling. It has high fibre content and hence has low digestibility. The protein is deficient in lysine, tryptophan and threonine. The sulphur containing amino acid is very high. The leusine to isoleusine ratio is generally unfavourable. The content of thiamine is 0.33 mg per gram.

Synthesis of starch

In the synthesis of starch an enzyme called phosphorylase is involved. There are two types of phosphorylase enzymes. The phosphorylase enzyme-'a' can be converted in to phosphorylase-'b' enzyme in the presence of enzyme phosphorylase phosphatase. In the above process 4 molecules of phosphate ions are removed and the molecular weight of the enzyme is reduced to half, again phosphorylase-'b' enzyme can be converted into phosphorylase-'a' in the presence of an enzyme phosphorylase-'b' kinase.

$$\text{Phosphorylase-a} \xrightarrow[\text{-4 Pi}]{\text{Phosphorylase phosphatase}} \text{Phosphorylase-b}$$

$$\xleftarrow[\text{+4 Pi} \quad Ca^{++}\text{ ions}]{\text{Phosphorylase-b kinase}}$$

In this process the molecular weight of the substance is doubled i.e. the four phosphate ions are incorporated. The above reaction takes place in the presence of an activator i.e., Ca^{++} ions .

In the synthesis of starch three well defined mechanisms have been identified and in each mechanism a different enzyme is involved. They are as follows:-

i. Glucose-1-phosphate

In this mechanism the enzyme involved is called phosphorylase. The starch is formed through a glucose transfer or glucosyl transfer from an energy rich glucose-1-phosphate in the presence of an acceptor and an enzyme.

$$\text{Glu-1-}PO_4 + \alpha\text{.D.Glucopyronosyl (1+4)} \xrightarrow[\eta\text{ (acceptor)}]{\text{Phosphorylase}} \alpha\text{.D.Glucophyronosyl }(\eta+1) + \text{Pi}$$

ii. Nucleoside disphosphate mechanism

Here the mechanism is called glycogen biosynthesis involving the UDPG and in the presence of enzyme called UDPG glycogen transglucosylase

$$\underset{\text{(Donor)}}{\text{UDPG}} + \text{Acceptor (G)}\eta \xrightarrow[\text{transglucosylase}]{\text{UDPG glycogen}} \text{UDP + Starch (G)}\eta$$

The above mechanism takes place in the most of the plant system. Later, scientist have identified ADPG as donor in higher plants and further stated that ADPG is more efficient than UDPG in the biosynthesis of starch.

iii. Sucrose

In some plants especially in cereals the carbohydrates are translocated in the form of sucrose which is an essential product of photosynthesis. The mechanism of sucrose-starch conversion is crucial one and in the conversion two pathways are identified.

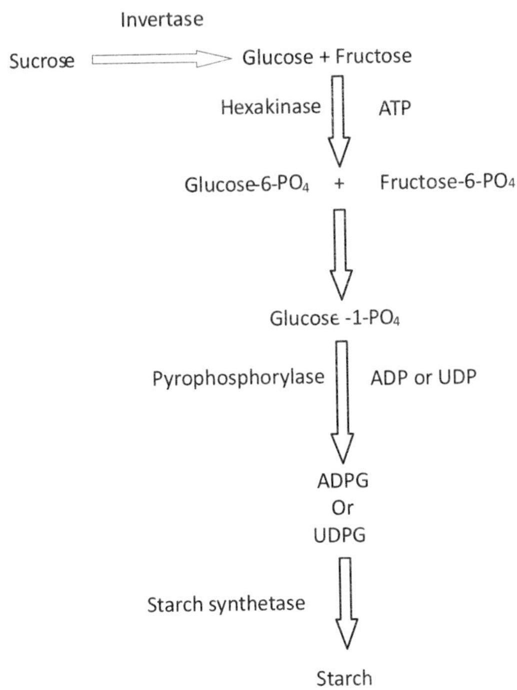

The essential part of the above mechanism involves the conversion of sucrose into glucose and fructose by the enzyme invertase. Then the glucose and fructose are converted into glucose-6- PO_4 by another enzyme called hexakinase in the presence of ATP. Again these are converted either into ADPG or UDPG by the enzyme phosphorylase. Finally it is converted into starch by the action of enzyme the starch synthetase.

Sucrose is directly converted into ADPG or UDPG in the presence of an enzyme sucrose synthetase and then into starch by another enzyme called starch synthetase.

Chapter 3

Nutritional Quality of Sugarcane

Sugarcane is a commercial crop. The quality of cane juice plays a vital role in sugar recovery.

(i)	Water	-	69 to 75%
(ii)	Sucrose	-	8 to 16%
(iii)	Reducing sugar	-	0.2 to 2%
(iv)	Organic compounds	-	0.5 to 0.6%
(v)	Organic matter	-	0.5 to 1%
(vi)	Nitrogenous compounds	-	0.5 to 1%
(vii)	Ash content	-	0.3 to 0.5%
(viii)	Fibre content	-	10 to 16%

Composition of sugarcane juice

(i)	Water	-	75 to 85%
(ii)	Sucrose	-	10 to 21%
(iii)	Reducing sugars	-	0.3 to 3%
(iv)	Organic compounds	-	0.5 to 1%

Sucrose

Sucrose is the most commonly found disaccharide in nature and its distribution is universal among the photosynthetic plants. Some of the crops like sugarcane and sugar beet contain higher amount of sugar. The basic unit of sucrose is nothing but the combination of glucose and fructose with the removal of one molecule of water. In the structure of sucrose there is no potentially active carbonyl group and it will not reduce the Fehling's solution. On hydrolysis of sucrose it will give rise to equal amounts of glucose and fructose which will reduce the cupric ions to cuprous ions. The hydrolysis of sucrose is very easy when compared to other disaccharides and the process is called ***inversion*** of sucrose and the sugar so formed is called ***invert sugar***.

Sugarcane is composed of fibrous cellular skeleton frame work holding canes. The fibrous matter left behind after the extraction of juice is called ***beggassee***. The juice volume varies depending on the method of extraction. The fibrous content is also variable. It may be as low as 10% with softcanes (Chewing canes) and 12 to 14% with hard canes (industrial canes). The central pithy portion of the internode contain less of fibre and is surcharged with higher amounts of juice when compared to rind fibre, less juice and more non-sugary impurities. Juice of soft canes with long internodes is of high purity. The juice of short internode canes have more impurity comparatively. Cane sugar is called ***sucrose***. It is sweet to taste, crystalline and non-hygroscopic. Presence of reducing sugars, acids and mineral matter in the juice will affect the crystallization of sucrose to some extent. Sucrose content of cane depends on variety and may range from ***13 to 22%***. Based on sucrose content the canes are classified as follows:-

(i) 12 to 15% sucrose - Poor canes

(ii) 16 to 19% sucrose - Fairly rich canes

(iii) > 20% Sucrose - rich canes

The quantity of total solid in a cane is estimated using a special type of hydrometer called ***brix meter***. It has been calibrated in such a way that the total solids percent can be read in the juice as Brix. The total solids in the juice of young and immature cane may be as low as 15%. It increases as the cane matures and goes upto 22% in the case of rich canes. The ratio of sucrose to total solid is expressed as percentage and is called the coefficient of purity

$$\text{Coefficient of purity} = \frac{\text{Sucrose content}}{\text{Brix}} \times 100$$

The maturity of cane depends on

1. Sucrose content
2. Brix
3. Purity

Matured cane should have 20 brix, 16 to 18% sucrose and 85 to 95% purity.

Another method of determining the maturity of cane is top-bottom ratio. It is a rough and simple method of determining the maturity of cane. It is the ratio of brix of the juice between the top half of the cane to bottom half of the cane. If the ratio is one, cane is said to be matured. The other sugars present in cane juice are dextrose and levose which are reducing sugars and are commonly called as glucose. They are mildly sweet, non-crystalline and hygroscopic in nature. Comparatively levose is more hygroscopic and it induces softness in jaggery. When the glucose content increases jaggery does not set properly. As the cane matures the glucose content decreases rapidly and may be reduced to 0.5% and converted into sucrose when the cane is said to be matured. The transformation of sucrose back to glucose within sugarcane or in juice is referred to as inversion. It takes place in the crop that leads to the formation of fresh roots and shoots from the node. When the crop is kept in flood, there is a slow inversion called over ripening. The mineral matter in sugarcane mainly consists of PO_4, SO_4 and chlorites of Ca, K and Mg. The contents increase with in immature cane and decrease with mature cane. Their presence interferes with proper setting of jaggery. The chlorites in juice promotes inversion and impart salty taste to jaggery. Nitrogenous substances in juice have a deleterious effect on odour of jaggery. Nitrogen compounds like pectin and gums are present in juice as colloidal state and they interfere with clarification during boiling and proper setting of jaggery. Succinic, aconitic and oxalic acids constitute the main free acid in sugarcane.

Sugarcane juice is light greenish in colour due to the presence of chlorophyll which is insoluble. Anthocyanin constituting colouring matter of rind. Polyphenols or tannins present in rind and nodes also contribute to the colour of the juice. Tannins get darkened while coming into contact with iron rollers in crushing mills and oxidases present in juice acting on tannins impart a light brown colour to cane juice.

Biosynthesis of sucrose

It involves many complex factors. The synthesis of sucrose takes place from simple sugars which will fall on the leaf blade of sugarcane. For the synthesis of sucrose adequate temperature and aeration is essential. The sugarcane plant has been described as the most efficient for the utilization of the sun's energy, because of the C_4 pathway. Sucrose may constitute 80 to 90 per cent of the total photosynthate within one hour. The transport of sugars from leaves takes place with a high speed of 2.5 cm/min. Biosynthesis of sucrose in sugarcane takes place through two pathways.

1. In the first case the glucose-1- PO_4 combines with the fructose to form sucrose and PO_4 ions.

   $$\text{Glucose-1-PO}_4 + \text{Fructose} \longrightarrow \text{Sucrose} + \text{PO}_4$$

   In this process the glucose-1-PO_4 acts as donor.

2. In the second case, the UDPG that is present in the leaf blade of sugarcane combines with the fructose in the presence of an enzyme sucrose phosphate synthetase. It will give rise to UDP + Sucrose-6-PO_4. This sucrose-6-PO_4 is again converted into sucrose + PO_4 ions.

   $$\text{UDPG} + \text{Fructose-6-PO}_4 \xrightarrow[\text{Synthetase}]{\text{Sucrose PO}_4} \text{UDP} + \text{Sucrose-6-PO}_4$$

   $$\text{Sucrose-6-PO}_4 \longrightarrow \text{Sucrose} + \text{PO}_4$$

   Sometimes the UDPG will directly combines with the fructose in the presence of the enzyme sucrose synthetase. It will give rise to UDP and sucrose.

   $$\text{UDPG} + \text{Fructose} \xrightarrow[\text{Synthetase}]{\text{Sucrose PO}_4} \text{UDP} + \text{Sucrose}$$

❑❑❑

Chapter 4

Nutritional Quality of Pulses

Nutritive value of Pulses

In India, people are mostly vegetarian, depending largely on cereals and pulses, as their staple food, which provide the main source of dietary proteins and calories. The proteins from pulses are known to be of inferior quality, due to the deficiency of sulphur-containing amino acids as well as due to other factors like digestibility of amino acids, anti-nutritional factors, etc.

Pulses are commonly consumed in combination with cereals and make good the deficiency of lysine in cereals, while cereals supply sufficient sulphur-containing amino-acids to supplement the deficiency of pulse protein. In general, pulses contain 20 to 30 per cent protein, about 60 per cent carbohydrates, 1.0 to 2.5 per cent fat, and are fairly good sources of thiamine, nicotinic acid, calcium and iron. Pulses carbohydrates (mainly starch), are well absorbed and utilized, providing equivalent energy value in terms of calories on unit weight-basis as that of cereals.

Pulses	Ash %	Protein %	Fat %	Carbhoyrate %	Curde fibre %
Red gram	3.8	17.6	1.3	61.4	5.8
Black gram	7.1	19.3	0.7	62.4	0.5
Green gram	4.1	21.2	1.1	57.6	3.8
Lab-lab	2.2	19.7	1.1	56.5	4.5
Cowpea	3.7	24.1	1.0	54.5	3.8
Soyabean	4.6	43.2	19.5	20.9	3.7

Studies on the nutrient composition of seed parts of bengalgram, greengram, redgram, pea, lentil, cowpea and soyabean revealed that cotyledon contained about 90 per cent of protein and minerals of the whole seed, and constituted the principal component, mainly accounting for the food value of the whole seed. Seed coat accounted for 32 to 50% of Ca of the whole seed.

Proteins and amino acids

Legume protein contain roughly 70% globulins, 15 to 20% albumins and 15 to 20% glutelins. Globulins are storage protein and main protein constituent in seeds. Albumins and glutelins are non-storage proteins. They are present in membranes. Storage globulin in red gram is called ***cajanin and con-cajanin.*** Variation in protein content occurs due to environment and production practices. The protein of other pulses are mainly globulins.

Amino acids

There are two ways of synthesis of Amino acid:

1. Reductive amination ≺ Synthesis of glutamic acid / Synthesis of aspartic acid
2. Transamination ≺ Synthesis of aspartic acid / Synthesis of alanine

The pulse seeds are rich in amino acids. Amino acids of pulses are mostly lysine, isoleucine, leucine, threonine and valine but have a decreased nutritive value due to small amount of sulphur containing amino-acids. Red gram contains greater amount of tyrosine, cystine, arginine and lysine whereas bengalgram contains greater amount of arginine.

Lipids

They are present in small amount in the seed and generally it varies from 1 to 2% except chickpea which contains 6%. Among the pulses cowpea contained greater amount of total saturated fatty acids whereas blackgram and soyabean contained greater proportion of unsaturated fatty acids. Lipids influence the development of flavours in legume seeds. There is no cholesterol in pulses.

Carbohydrates

Combined with protein pulses are greater source of carbohydrate. Mainly comprises of water insoluble components and also starch and cellulose. Main sugar present is sucrose. They also contain oligosaccharides like raffinose, stachyose, verbascose etc. these are implicated in gaster intestinal trouble. They accumulate in the large intestine where natural gut microflora use these sugars and produce gas.

Minerals

Pulses are a good source of minerals. They are generally rich in Ca, Fe, P and K. All pulses are inferior to soybean with regard to Ca, Mg, P, Fe, Zn, Cu, etc. contents. Minerals in cooked legume is 1/2 to 1/3rd content of raw legume.

Vitamins

Mung bean seeds have greater source of thiamine, niacin, pantothenic acid and vitamin A. Vitamin content will be more in the sprouted seeds than in the raw seeds.

Red gram

Main protein present is globulin. Two isolates of globulin are cajanin (58%) and con-cajanin 8% of total N. These two differ in structure and tryptophan constituent. It has greater proportion of tyrosine and fairly rich in cysteine, arginine and lysine. Quality of red gram protein is inferior to Bengal gram protein.

Bengal gram

Globulins of Bengal gram constitutes 96% of protein. They are rich in arginine (19.8%) in addition to tyrosine and lysine but differ in cysteine and tryptophan contents. Contains modest amount of vitamin A, thiamine, little riboflavin and fair amount of niacin. Good source of Fe and Cu. Rich in Ca compared to major cereals.

Lab-lab

Major protein is globulin. It contains α and β globulins. α globulin is rich in tyrosine and lysine. β globulin is low in cystein than α globulin.

Soybean

Soybean protein is a good source of lysine content but is deficient in methionine.

Cereal grain and legume protein

1. In an average cereal seed contains 10% protein whereas legume seed contained 30% protein.
2. Major storage protein of most cereals are soluble prolamines but legumes contains salt soluble globulins.
3. Major storage proteins of most cereals contain very little lysine and tryptophan while storage proteins of legumes are low in methionine and cysteine.
4. Storage proteins of cereals occur in the endosperm and also as deposit in the endoplasmic reticulum. Synthesis of protein in legumes takes place in cotyledons and embryonic axis and deposited inside vascular membranes.

Biological evaluation of protein

Biological evaluation has been recognized as a good test criterion for assessing the quality of dietary proteins. Different pulses were found to vary widely in their biological parameters. Their digestibility coefficient, biological value and protein efficiency ratio varied from 60 to 90 per cent, 45 to 78 per cent and 0.7 to 1.1 per cent respectively.

□□□

Chapter5

Nutritional Quality of Oil Seed Crops

Chemistry of oil

The oils and fats are mixed triglyceride (esters of glycerol with fatty acids). Glycerol is a trihydric alcohol and is capable of reacting with 3 molecules of fatty acids.

Groundnuts

They have outer thick woody shell inside is normally embedded with 2/3rd seed called kernels. Seed consists of two cotyledons and a germ. It has an outer thin skin called testa. Colour of testa varies from red to brown, purple and white depending on type and variety. The weight of different parts are as follows.

Testa - 4 to 5% weight of kernel
Cotyledons - 92 to 94% weight of kernel
Germ - 3 to 4% weight of seed cotyledons

Testa has protective role against pests and diseases. Cotyledons act as storage organ and supply food during germination.

Chemical Composition

Constituents	Testa	Germ	Cotyledon
Moisture (%)	9.00	-	4 to 13
Protein (%)	11 to 13	26.5 to 27.8	21.4 to36.4
Fat (%)	0.5 to 1.9	39 to 43	36 to 54
Total carbohydrate (%)	48 to 52	-	6 to 24
Reducing sugars (%)	1 to 1.2	7 to 9	0.1 to 0.4
Sucrose (%)	-	12	1.9 to 6.4
Starch (%)	-	-	0.9 to 6.4
Crude fibre	21 to 34	1.6 to 1.8	1.6 to 1.9
Ash	1.9 to 4.6	1.9 to 3.2	1.8 to 3.1

Kernel

Moisture	-	5 to 7%
Fat	-	46 to 50%
Fibre	-	1 to 3%
Protein	-	26 to 31%
Carbohydrate	-	1 to 3%
Ash	-	2.5 to 3%

Germ

Contains higher amount of protein, reducing sugars and disaccharides.

Cotyledons

Higher amount of oil and protein.

Testa

1 to 5% phenolic compounds in addition to carbohydrate, cellulose and protein.

Groundnut oil

Oil content of groundnut varies from 40 to 54%. It depends on variety and to some extent on agro climatic condition. The oil is yellow in colour.

Characteristics of oil

Melting point	-	0 to 3°C
Iodine value	-	85–98
Acid value	-	2–6%
Saponification value	-	180 to 195
Free fatty acids	-	0.02 to 0.6%

Oil has greater proportion of oleic acid (47.9%) than linoleic acid (29.9%). Palmitic, stearic and arachidinic acids are present in lesser proportions. Ratio of oleic and linoleic acid has important bearing on shelf-life of edible oil. This value for groundnut oil is 1.6. At value around 2 will give better shelf-life.

Protein

It contains two types of globulins viz., arachin and canarachin. Lysine, methionine threonine, isoleucine are limiting amino acids in groundnut protein.

Cake

The residue left behind after the extraction of oil is called a cake.

Composition

Moisture	-	8 to 10%
Carbohydrate	-	22 to 30%
Crude protein	-	46 to 60%
Crude fibre	-	3.8 to 7.5%
Mineral matter	-	4 to 5.7%

In our country it is mainly used as cattle feed and as organic manure. The cake also contains 1 to 4% phosphoric acid and 1 to 2% potash.

Groundnut shell composition

Cellulose	-	68%
Protein	-	7%
Carbohydrate	-	21%
Lipids	-	1 to 2%
Minerals	-	4.5%

Shell content of pods varies from 20 to 40%. Spreading variety has higher proportion of shells than Bunch variety.

Haulms

Proteins	-	8 to 15%
Lipids	-	1.9 to 2.9%
Crude fibre	-	22 to 30%
Carbohydrate	-	38 to 47%
Moisture	-	7 to 10%
Minerals	-	9 to 17%

It is mainly used as cattle feed. Contains more protein, lipids and minerals than cereal fodder.

Mycotoxins

Groundnut fruit development takes place with close contact with soil. Many species of soil fungi have been isolated from healthy and rotten pods. Main fungi which produces mycotoxin in agricultural commodities are *Aspergillus, Fusarium* and *Penicillium*. One such fungus commonly found is *Aspergillus flavus* which produces aflotoxin when growing in groundnut seeds/groundnut products. Four mycotoxins viz., aflotoxin, citrinin, ochratoxin-A and Zeavalenone occur naturally in groundnut. Of which aflotoxin is toxic and carcinogenic to a wide range of animals and human beings.

Coconut

Kernel/endosperm is the valuable part. Percentage composition of kernel varies depending upon the variety and the environment in which it grows. The composition of various parts of coconut is as follows :

Constituents	Kernel of green coconut	Kernel of ripe coconut	Copra (dried)	Tender coconut water	Cocomilk	Toddy
Moisture (%)	75.90	36.30	6.80	95.50	52.00	84.4
Ash (%)	0.98	1.00	2.00	0.40	1.10	0.26
Crude fat(%)	6.43	41.60	63.70	0.10	27.00	0.04
Protein (%)	19.41	4.50	7.60	0.10	4.00	0.12
Fibre (%)	2.50	3.60	3.80	-	-	-
Carbohydrate (%)	4.76	13.00	16.00	4.00	15.90	15.20

Moisture, protein and ash content of endosperm decrease as nut matures whereas fat content increase. Moisture and protein contents of fresh endosperm are increased first in 8 months nut and slowly decrease in the subsequent stage of maturity. In case of protein, practically there was no change after ten months. Fat content reached its peak at 12-13 months which subsequently decreased until 15th month.

Coconut protein

Chief protein of coconut kernel is globulin. When coconut is processed into copra and oil in traditional way, major portion of protein passes into the coconut cake.

Mineral

It is a poor source of minerals

Calcium	-	0.01 mg/100g
Phosphorus	-	0.24 mg/100g
Iron	-	1.70 mg/100g
Copper	-	0.32 mg/100g
Sulphur	-	0.44 mg/100g

Vitamins

Good source of vitamins A, B, C and E. Vitamin C is highest in green nut with soft pulp. Vitamin C is maximum in germinating fruit. Fair amounts of vitamins B1, B2, A and C are present in toddy.

Copra

Dried endosperm of coconut. Highest source of fat. Contain 65 to 70% oil. Essential requirement of copra drying is to bring down moisture content of wet nut from 55 to 60% to 5 to 6%. Eleven to twelve months old nuts are utilized for copra making.

Coconut oil

It is colourless to brownish yellow in colour. It is fluid in warm tropical climate but changes to solid fat in temperate climate. Its melting point in solid state ranges between 23 to 26°C. The oil consists mainly of glycerides of lauric (47.32%), myristic (17%) and other acids. In solid state coconut oil is less susceptible to deterioration than in liquid. Its characteristics are as follows :

Specific gravity	-	0.978 at 25°C
Saponification value (No.)	-	251 to 263
Iodine value (No.)	-	8 to 9.6
Polenske (No.)	-	15 to 18

Chemical properties

High in saturated fatty acids particularly lauric and myristic with notable amount of still lower fatty acids. It is low in unsaturated and poly unsaturated acids especially linoleic acid (1.87%). It has the highest saponification value due to its low unsaturated fatty acid content. On account of its low iodine value it is classified as non-drying oil.

Coconut cake

Forms about 32 to 40% of coconut crush. Mainly used as ruminant feed. Composition of cake varies with the method of oil extraction.

S.No.	Constituent	Expelled cake	Solvent extracted cake
1.	Moisture (%)	7.9	8.9
2.	Fat (%)	6.7	2.4
3.	Protein (%)	21.2	21.4
4.	Carbohydrate (%)	47.4	47.4
5.	Fibre (%)	11.2	13.3
6.	Ash (%)	6.5	6.5

Compared to other oil cakes it is poor in protein but rich in fat. It is also used as organic manure for field crops but it has the highest C:N ratio and therefore not considered for manuring seasonal crop.

Carbohydrate

Derived copra contains 20% of carbohydrate comprising mainly sucrose and cellulose. Reducing sugars found in milk of tender coconut. Sucrose is present in the milk of matured coconut. Coir/coconut fibre belongs to a group of hard structural fibre. It is an important commercial product obtained from husk. Husk constitutes 35 to 45% by weight of whole nut when derived. Percentage of husk decreases with ripening of nut. These are two types of coir fibre.

Brown fibre

Extracted from ripe dry husk by mechanical defiltering process. There are two types.

Bristle type - Used for brush making

Mattress type - Mattress making

White fibre

Extracted from green husk by bacteriological process of rotting (by immersing green husk in water).

Coirpith or coir dust

Waste product in coir industry constitute as much as 70% of husk. One of the suggested use of coir dust is manure. It can hold light times the weight of H_2O. The composition is as follows:

Moisture	-	11.9% (air dry basis)
Ash	-	8.7%
Cellulose	-	35%
Fat	-	1.8%
Lignin	-	35.7%
Pentosan	-	11.95%

It decomposes very slowly in soil.

Oil palm

The highest oil producer per unit area.It produces two kinds of oils.

- Palm oil
- Palm kernel oil

Palm oil

It is obtained from fleshy mesocarp of fruit which contain 45 to 50% oil. Light yellow to orange red in colour. Depth of colour depends on amount of carotenoid content. Oil content increase is proportional to saturated palmitic acid. It also contains considerable quantities of oleic and linoleic acid. Poorly prepared oil has a higher free fatty acid caused due to catalytic action of lipase in unsterilised and bluish fruits. This renders it unsuitable for edible purpose. It is used for the manufacture of soaps and candles.

Palm kernel oil

It is obtained from the kernel. The endosperm contain 50% oil. It is a hard oil resembling coconut oil. It is readily interchangeable with coconut oil. It has higher proportion of saturated acids predominantly lauric acid. Solid at room temperature and colourless. Used in edible fats, in confectionaries, bakeries, ice-cream, soap powder, detergent toilet soap. The cake is used as an important livestock feed. In addition to supply of fat, palm is rich source of carotene especially α carotene the richest precursor of vitamin A.

Sesamum

It is a semidrying oil. It is a substitute for olive oil. It is used as cooking oil. Poor grade is utilized in the manufacture of paints and soaps. Seed contain 45 to 55% oil. It is odourless. It is not liable to become rancid. Main unsaturated fatty acids present are oleic and linoleic with 40% each and 14 per cent of saturated fatty acids. Linolenic acid is not present. Sesamum seed is important for human nutrition on account of higher content of sulphur containing amino acids especially methionine. Main limitation of using this seed is presence of cuticle which contribute dark colour and bitter taste to the seed. It has high crude fibre content and also contain calcium oxalate.

Sunflower

The oil is pale yellow salad oil. It is also semidrying oil. Used in blend with linseed and other drying oil in paints and varnishes. It also used as lubricant. The hull constitute 35 to 50% seed. Used as fodder in feed cakes and as bedding material for livestock. Seed contains 25 to 35% semidrying oil. Linoleic acid is present to an extent of 44 to 72%. It depends on the condition in which the crop is grown. Slow maturing seeds and crops grown in cool climate is highest in linoleic acid and less in oleic acid. Protein content of seed ranges from 14 to 15% in stem and husk are rich in K. Sunflower oil is non cholesterol in nature and hence fetches highest premium in the market.

Physical and chemical properties of oils and fats

Physical properties

State

Fats containing saturated fatty acids are solid at ordinary room temperature. The animal fats belong to this category. Most of the plant fats on the contrary possess unsaturated fatty acids and hence liquid at room temperature.

Colour, odour and taste

When pure the fats are colourless, virtually odourless and possess an extremely blunt taste. They are capable of absorbing a variety of odours and hence have flavour during storage.

Solubility

The fats are only sparingly soluble in water. They are hydrophobic in contrast to hydrophilic substances like many carbohydrates and proteins. However fats are freely soluble in organic solvents like chloroform, ether, acetone and benzene. These solvents are called "fat solvents". The solubility of fatty acids in organic solvents infact decreases with the increase of chain length.

Melting point

It depends on the chain length of the constituent fatty acid and the degree of unsaturation. Greater the degree of unsaturation of the constituent fatty acid, the lower is the melting point of the fat.

Specific gravity

The specific gravity of the fats is < 1.00 (about 0.86) and therefore, they float on water surface. Solid fats are lighter than liquid fats.

Emulsification

It is the process by which a lipid mass is converted into a number of small lipoid droplets. The fats may be emulsified by shaking either with water or with emulsifying agents like soaps, gums, proteins etc. An emulsifying agent helps in the production of a finely divided suspension of a fat in an aqueous medium. The process of emulsification is of great metabolic significance. In fact, the fats have to be emulsified before they can be absorbed by the intestinal wall. The process is accomplished by the bile juice secreted from liver.

Surface tension

When liquid fat is poured on water it spreads uniformly over the surface of water in the form of a unimolecular layer and thus reduces the surface tension of water.

Chemical properties

Hydrolysis

The fats are hydrolysed by the enzymes lipases to yield fatty acids and glycerol. The lipases catalyze the reaction at a slightly alkaline pH 7.5 to 8.5 in a stepwise manner. The fats first split and produce diglycerides, part of these are then split into monoglycerides, finally part of the monoglycerides split to yield fatty acid and glycerol.

There are three distinct steps in the synthesis of fats and oils:

1. Synthesis of glycerol.
2. Synthesis of fatty acids.
3. Condensation of glycerol and fatty acids.

H_2C – OOCR1		H_2C – OOCR1
HC – OOCR2	→	HC – OOCR2 + R3COOH
H_2C – OOCR3 + HOH		H_2C – OH
Triglyceride Water		Diglyceride

H_2C – OOCR1		H_2C – OOCR1
HC – OOCR2	→	HC – OH + R_2COOH
H_2C – OH + HOH		H_2C – OH
Diglyceride Water		Monoglyceride

H_2C – OOCR1 + HOH		CH_2OH
HC – OH	→	CHOH + R_1COOH
H_2C – OH		CH_2OH
Monoglyceride		Glycerol

Saponification

The hydrolysis of fats by alkali is called saponification. This reaction results in the formation of glycerol and salts of fatty acids which are called soaps. The soaps are of two types, hard and soft. Hard soaps are the sodium salts of the higher fatty acids. Soft soaps are the potassium salts of higher fatty acids and are marketed as semi solids. Calcium soaps are used industrially for lubricating greases. Zinc soaps are employed in the manufacture of talcum powder and other cosmetics. Lead and magnesium soaps are used in paint industry to hasten the process of drying.

$$\begin{array}{lcl} H_2C-OOC-C_{17}H_{33} & & CH_2OH \\ HC-OOC_{17}H_{33} & \longrightarrow & CHOH + 3C_{17}H_{33}COO\,Na \\ H_2C-OOC_{17}H_{33} + 3\,Na\,OH & & CH_2OH \\ \text{Triolein} & & \text{Glycerol Na oleate (soap)} \end{array}$$

Oxidation

Fats and oils containing unsaturated fatty acids absorbs oxygen at their double bonds in the molecule and then gets saturated. This type of oxidation is called autooxidation.

Rancidity

When fats and oils are exposed to light, air, heat or moisture for a long time they develop off flavour. Then the oil or fat will produce a bad smell and the off flavour is known as rancidity. The rancidity is of two types.

(a) Hydrolytic rancidity

It is produced as a result of hydrolysis of fats and oils with the liberation of small amount of one or more volatile fatty acids which are responsible for the off flavour.

(b) Oxidative rancidity

The unsaturated fatty acid fragments are oxidized at the double bond with production of aldehydes, ketones and acids with fewer carbon atoms in the molecules that are responsible for the bad odour mainly due to the decomposition products.

Hydrogenation

It is a process in which the hydrogen atom is combined with unsaturated fatty acid. The process is completed in the presence of a catalyst. This process is done in order to prevent the oxidative changes of fats and oils and also to increase the keeping quality of oil.

$$\text{Unsaturated fatty acid} + O \xrightarrow[\text{Catalyst}]{H} \text{Vanaspathi}$$

It is the main reaction of the carboxyl group of the fatty acids in which a molecule of the acid and alcohol react to produce one molecular of ester and water.

Fats

Fats are defined as the esters of glycerol and fatty acids or as the triglycerides of fatty acid. Fatty acids are long chain organic acids having usually from 4 to 30 carbon atoms, they have a single carboxyl group and a long non-polar hydrocarbon tail which gives most lipids their hydrophobic and only greasy nature.

Types

- Saturated fatty acid - Containing only single bond. *Eg.* Lauric and palmitic.
- Unsaturated fatty acid – Presence of double or triple bonds. *Eg.* oleic, linoleic and linoleinic.
- Hydroxy acid – May have hydroxyl group in the chain. *Eg.* ricinoleic acid.
- Cyclic fatty acid – Possess ring structure. *Eg.* Sterculic acid.

❑❑❑

Chapter6

Analytical Constant of Fats and Oils

Analytical constants

The chemical nature of fatty acids and the number of hydroxyl groups present in the fat molecule involve various analytical tests. These are called analytical constants or chemical constants. They are

1. Acid value

It is the number of milligrams of KOH required to neutralize the free acids present in 1 gm of fat or oil. The acid value indicates the quantity of free fatty acid present in fat. Obviously, a processed and properly stored fat has a very low acid number. The difference between the saponification number and acid number is known as the ester number.

2. Saponification number

It is the number of milligrams of KOH required to saponify 1 gm of fat or oil. The saponification number thus provides the information on the average chain length of the fatty acids. It inversely related with the chain length of the fatty acids. The shorter the average chain length of the fatty acids the higher is the saponification number.

It is an index of the mean molecular weight of the fatty acid glycerides comprising fat and to determine the admixture of foreign fat with butter fat. Butter fat and coconut oil have large quantities of this short chain fatty acid glycerides.

3. Iodine number

It is the number of grams of iodine absorbed by 100 g of fat or oil. The iodine number is thus, a measure of the degree of unsaturation of the fatty acids in the fat or oil. Oils like soybean, corn and cotton seed have higher iodine numbers (133, 127 and 109 respectively) than the solid fats such as beef fat, because the former possesses more unsaturated fatty acids in the fat molecule. However, the iodine number gives no indication of the number of double bonds present in the fatty acid molecule. It is a measure of the unsaturated linkages present in the fat or oil.

Polenske number

It is the number of milliliters of 0.1 N KOH required to neutralize the steam volatile fatty acids insoluble in water obtained from 5g of a saponified fat or oil. It is of special importance in connection with the analysis of butter and oleomargarines.

Reichert-Meissl number (RM number)

It is the number of milli litres of 0.1 N KOH required to neutralize the steam volatile water soluble fatty acids derived from 5 gm of a sponified fat or oil. The RM number increased with the quantity of short chain fatty acids in the fat molecule. The RM number of coconut and palm oil range between 5 and 8. Butter fat is exceptional in having a high RM number ranging from 17 to 35. This high value makes possible the defection of any foreign fats which are sometimes adulterated in the manufacture of butter. Butter fat contains butyric acid which is volatile and soluble in water.

Acetyl number

It is the number of milligrams of KOH required to neutralize the acetic acid obtained by saponification of 1 gm of fat or oil. After it has been acetylated (the treatment of fat of fatty acid mixture with acetic anhydride results in acetylation of all alcoholic OH groups). The acetyl number is thus, a measure of the number of OH groups present in the fat form. Example, the castor oil has a high acetyl number (142) because of its high content of ricinoleic acid.

Analytical constants of oils and fats

Name oil/fat	Acid value	Saponification value	Iodine value	R.M. value	Polenske value
Coconut oil	5-13	250-264	8.0-9.5	6.0 to 9.0	15 to 20
Sesame oil	1to10	188-193	103-107	1.0 to 1.2	-
Castor	0.3to40	178 to 188	80 to88	0.2 to 0.2	-
Linseed	1 to 8	190 to 196	170to20	30.2 to 0.3	-
Groundnut oil	2 to 6	186 to 196	83 to 105	0.4 to 0.5	-
Butter	0.4 to2.0	220 to 241	26 to 38	24 to 32	1.5 to 3.0
Ghee (Cow)	-	223 to 230	32 to 45	21 to 34	0.7 to 2.0
Ghee (Buffalo)	-	229 to 236	27 to 44	27 to 36	0.8 to 2.2

Nutritive value of fats and oils

The lipids present in the diet, animal and human body include the following :

- Triglyceride
- Phospholipids
- Waxes
- Sterols

Generally the plant fats are relatively rich in unsaturated fatty acids especially poly unsaturated acids. They are liquid at room temperature. These have usually high iodine number. They have low RM number. Stored mainly in seeds and fruits. Oxidative rancidity is observed less frequently. Eg. Olive oil, castor oil, soybean oil, corn oil.

Poly unsaturated fatty acids are also known as essential fatty acids. They are found to be essential for the growth. The two major groups of EFA are linoleic and linolenic acids.

Rich sources of Essential Fatty Acids

Safflower, sunflower, corn, sesame and soybean oils are rich sources (40 to 78%) of EFA. These oils must form atleast 50% of the fat in the diet.

Good sources

Rape seed, peanut and rice bran oils (27 to 32%).

Fair sources

Egg yolk, poultry fat and pig body fat (15 to 22%).

Poor sources

Coconut oil, palm kernel oil, vanaspathi (hydrogenated groundnut oil), butter and ghee, beef and mutton fat are poor sources of EFA (3 to 5%).

Toxic effects of some fatty acids

Erucic acid

It occurs to an extent of 45% in rape seed and mustard oils. These two oils are consumed extensively in human diets in some parts of

India, Sweden and Canada. It has been reported that feeding diets containing rapeseed oil at a level of 20 to 50% causes myocardial fibrosis in the right ventricle in experiment animals.

Sterculic acid

It occurs in small amounts (0.6 to 1.2%) in cotton seed oil and in large amounts (70%) in *Sterculia foefuda* oil. The toxic effects of sterculic acid has been attributed to

1. Inhibition of desaturation of stearic to oleic acid and
2. Increased permeability of membranes leading to the development of "pink white" disease in chicks.

Ricinine

An alkaloid present in castor oil. It's a product of nitrogen metabolism. During flowering and fruiting this takes place. It is a monocyclic derivative of pyridine with a cyanidine. Succinine and glycerine are precursor of ricinine. Synthesis of ricinine in castor occur with the participation of lysine. Alkaloid accumulates in 28 days old plant. Concentration of ricinine increase from kernel towards capsule shell. The level of accumulation depends on protein and the quantity of oil.

❑❑❑

Chapter 7

Nutritional Quality of Fruits I

Mango

It is acknowledged as the king of fruits of this country. The unripe green mangoes are reported to have :

Moisture	-	90%
Protein	-	0.7%
Fat	-	0.1%
Carbohydrates	-	8.8%
Calcium	-	0.01%
Phosphorus	-	0.02%
Iron	-	4.5 mg/100g
Carotene	-	150 i.u.
Riboflavin	-	30 mg/100g
Ascorbic acid	-	3 mg/100g

Ripe fruits (on fresh weight basis)

Moisture	-	79 to 83%
Total soluble solids	-	13 to 21%
Total sugars	-	10 to 17%
Non reducing sugars	-	7 to 13%
Ash content	-	0.5 to 0.6%
Crude protein	-	0.4 to 0.6%

The sugars in mango comprise sucrose, glucose, fructose and maltose. Essential amino acids like alanine, aspartic acid, lysine, leucine, cystine, valine, arginine, phenylalanine and methionine are present.

Small amount of tannin is also present in the flesh (0.16%) and skin (0.1%) which is responsible for astringency. From the nutritional point of view the mango is a rich source of vitamin A, almost as rich as butter. Also has fair amount of vitamin C. The β-carotene (provitamin A) increase with ripening, whereas vitamin C registers a sharp fall on ripening. Green mangoes contain folic acid (Vit B) to an extent of 3.6 μg/100g.

Banana

It is a rich source of energy. The composition of fruit is given below :

Moisture	-	70 per cent
Carbohydrate	-	27 per cent
Crude fibre	-	0.5 per cent
Protein	-	1.2 per cent
Fat	-	0.3 per cent
Ash	-	0.9 per cent
P	-	290 ppm
Ca	-	80 ppm
Fe	-	6 ppm
β carotene	-	0.5 ppm
Riboflavin	-	0.5 ppm
Niacin	-	7 ppm
Ascorbic acid	-	120 ppm

It contains nearly all the essential nutrients including minerals and vitamins and has several medicinal properties. About 24 bananas each weighing around 100g would provide the energy requirement (2400 Cal/day) of a sedentary man.

Pineapple

It is a good source of vitamins A and B and is rich in vitamin 'C' and calcium. It also contains phosphorus and iron. It contains enzyme, bromelin and pineapple leaf is a good source of chlorophyll. The 3-methyl propionate esters comprise a significant fraction of the pineapple volatile components and have been adapted for use in pineapple flavours. Several odorous lactones have also been found in pineapple.

A pineapple fruit contains :

Moisture	-	85%
Sugar	-	13%
Protein	-	0.6%
Mineral matter	-	0.05%
Crude fibre	-	0.3%
Calcium	-	0.02%
Phosphorus	-	0.01%
Iron	-	0.9%
Acids	-	0.6%
Vitamin A	-	60 I.U./100 gm
Riboflavin	-	120 mg/100g
Vitamin C	-	63 mg/100g

Dried waste after juice extraction is a valuable cattle feed. Other by-products are alcohol, calcium citrate, citric acid and vinegar.

Papaya

They are used in the preparation of jam, soft drinks, ice-cream flavouring crystallized fruits and in syrup. The seeds are also used for their medicinal value. Papain prepared from the dried latex of immature fruits, is a proteolytic enzyme similar in action to pepsin and is used in meat-tendering preparation, in manufacture of chewing-gum; for degumming natural silk and to give shrink resistance to wool. It is a very wholesome fruit. It is second only to mango as a source of the precursor of vitamin A. The yellow pigment in the papaya is not carotene but caricaxanthine.

Composition

Moisture	-	90%
Protein	-	0.5%
Fat	-	0.1%
Carbohydrate	-	0.5%
Calcium	-	0.01%

Phosphorus	-	0.01%
Iron	-	0.4%
Vitamin S	-	2020 IU/100 g
Vitamin B_2	-	0.04 mg/100g
Vitamin C	-	40 mg/100g
Nicotinic acid	-	0.2 mg/100g
Riboflavin	-	250 mg/100g

Grapes Composition

Moisture	-	70 to 80%
Carbohydrates	-	15 to 25%
Glucose	-	8 to 13%
Fructose	-	7 to 12%
Pentoses	-	0.01 to 0.1%
Pectin	-	0.02 to 0.1%
Inosital	-	0.02 to 0.08%
Organic acids	-	0.3 to 1.5%
Tartaric	-	0.2 to 1.0%
Malic	-	0.1 to 0.8%
Citric	-	0.01 to 0.1%
Nitrogenous compounds	-	0.03 to 0.17%
Protein	-	0.01%
Amino	-	0.11%
Humin	-	0.002%
Amide	-	0.004%
Ammonia	-	0.012%
Tannins	-	0.1%
Mineral compounds	-	0.6%
Ca	-	0.025%
Cl	-	0.01%
Mg	-	0.025%
K	-	0.25%
P	-	0.05%
SO_4	-	0.03%

Sugars

The sugars of grape are glucose and fructose, generally accounting for 99 per cent crushed grapes and from 12 to 27 per cent or more of the weight of the mature fruits. Other sugars such as sucrose, raffinose, stachyose, melibiose, maltose and galactose are present in small amount. The main sugar translocated from the leaves to the fruit is sucrose. Most of the sugars in the berries are manufactured in the leaves. Another source of sugars in grape berries is from transformation of organic acids.

Acids

Tartatic and malic acids constitute 90% or more of total acidity. It is now well known that both leaves and fruits are capable of synthesizing malic and tartaric acids.

Tannins

These are complex esters of phenolic acids and sugars. These occur primarily in skin, stem and seed of grape. Its content influences the palatability of grapes and their products. The tannins give an astringent taste. Total phenols as gallic acid to an extent of 3770 mg/kg occurs in mature berries.

Nitrogenous compounds

Total N in crushed grape ranges from 10 to 200mg per 100ml. NH_4 cations and organic compounds such as amino acids, amines, peptides, nucleic acids and proteins constitute the major part of nitrogenous compound in grape berries.

Minerals

The mineral contents of grapes which constitute 0.2 to 0.6 per cent of fresh fruit weight are taken from soil.

Vitamins

Fresh grapes contain many vitamins. Fairly good amount of vitamin A is present which is retained in dehydrated grapes. Natural raisins contain no vitamin A. The B complex vitamins of grapes are thiamine, riboflavin, pyridoxine pantothenic acid, nicotinic acid, inositol, biotin and folic acid. Vitamin C content is 13.5 mg/100 ml of juice.

Pectin

These are derivatives of polygalacturonic acid. During ripening, the protopectin is transformed to pectin and the berries soften as a result of removal of middle lamella pectate. Vinifera grapes contain smaller amount of pectin compared with American grapes which form stable jelly.

Colour

The pigments of grapes are anthocyanidins modified by attachment of a molecule of glucose. There are five anthocyanins-cyanidin, delphinidin, petunidin and malvidin-that make up basic part of grape pigment. The yellow pigment of both white and red grapes, is quercetin-a flavone - and its glycoside, quercitrin-a flavanol. The anthocyanin content varies from 2500 to 3000 mg/kg. It should always kept in mind that it is the brightness, not the density of colour that makes table grapes more attractive.

Odorous constituents

During ripening grapes has a special aroma for a particular cultivar. The substance responsible for this has been isolated largely confined to the skin of berries and are synthesized in the berries itself. However, the precursor materials are produced in the leaves.

Flavour

It is the complex reaction of taste and olfactory receptors. Many substances contribute to flavour of which sugar, acids and tannins are dominant. Their relative amount and not the total amount determine the flavour.

Enzymes

Polyphenoloxidase is the main oxidizing system of grapes which is mainly confined to the skin. Other enzymes, such as phenolase, phosphatase, proteinase and sucrose are also found in skins and to lesser extent in pulp.

Guava

It is a rich source of vitamin C and pectin. Its composition is as follows :

Moisture	-	83%
Acid	-	2.45%
Reducing sugar	-	4.45%
Non-reducing sugar	-	5.23%
Total soluble solid	-	9.73
Ash	-	0.48%
Vitamin C	-	260 mg/100g

It is a fair source of vitamin A but a good source of both thiamine and riboflavin. It contains much iron but 80% of this is in the seeds and not utilizable. It is also a good source of calcium and phosphorus. Pink fleshed cultivars are poor in vitamin C content than the white fleshed ones.

Sapota

They are a good source of sugar which ranges between 12 and 14 per cent. The composition is as follows :

Moisture	-	74%
Carbohydrates	-	21.4%
Protein	-	0.7%
Fat	-	1.1%
Calcium	-	28mg/100g
Phosphorus	-	27 mg/100g
Iron	-	2 mg/100g
Ascorbic acid	-	6 mg/100g

The fruit skin is richer than the pulp in nutritive value.

□□□

Chapter 8

Nutritional Quality of Fruits II

Citrus

Different species of citrus fruits have different chemical composition. While in the sweet group, the principal constituents of the edible portions are sugars (glucose and sucrose) and acids (primarily citric acid and little of malic acid), the fruits of acid groups contain primarily the acids in the fruit juice. The rind of citrus fruits is rich in pectin and certain essential oils. The rind also contains certain glucosides (hesperidin in oranges, lemons and naringins in the grape fruit). It contains considerable amounts of ascorbic acid. The total soluble solids (TSS) in the fruit juice, in most of the sweet group of citrus, varies from 8 to 12%, while the titrable acidity usually ranges from 0.5 to 1.5 per cent. Usually, for oranges and mandarins a TSS-acid ratio of 8:1 and in grape fruit the ratio of 6:1 was considered to be optimum for marketable. In the lemons and limes, titrable acidity of fruit juice usually ranges from 5 to 6 per cent. The vitamin C content varies from 25 to 85 mg/100 ml of juice.

Pomegranate

Composition is as follows :

Moisture	-	78%
Protein	-	1.6%
Fat	-	0.1%
Fibre	-	5.1%
Carbohydrates	-	14.5%
Mineral matter	-	0.7%
Calcium	-	10mg/100g

Magnesium	-	12mg/100g
Phosphorus	-	70mg/100g
Iron	-	0.3mg/100g
Oxalic acid	-	14mg/100g
Thiamine	-	0.06mg/100g
Riboflavin	-	0.1mg/100g
Nicotinic acid	-	0.3mg/100g
Vitamin C	-	14mg/100g

The fruit juice easily ferments and may be used for the production of wine. The bark of the stem and root contain a number of alkaloid belonging to the pyridine group. Seed contains about 15% oil. The seed also contain 1.09mg vestrine/100g and 0.036mg coumestrol/100g of seeds.

Apple

Carbohydrate	-	13.4g/100g of edible protein
Protein	-	0.2g/100g of edible protein
Fat	-	0.5g/100g of edible protein
Minerals	-	0.3g/100g of edible protein
Fibre	-	1g/100g of edible protein
Vitamin C	-	1mg/100g of edible protein

Flavouring constituent in apple is ethyl-2-methyl butyrate. Aminoacids like aspartic and glutamic acids and alanine are present in apple. It also contains malic acid.

Pear

Carbohydrate	-	11.9g/100g of edible protein
Protein	-	0.6g/100g of edible protein
Crude fat	-	0.6g/100g of edible protein
Minerals	-	0.3g/100g of edible protein
Fibre	-	1g/100g of edible protein
Carotene	-	28 μg/100g of edible protein

Plum

Carbohydrate	-	11.1g/100 gm
Protein	-	0.7g/100 gm
Crude fat	-	0.5g/100 gm
Minerals	-	0.4g/100 gm
Fibre	-	0.4g/100 gm
Carotene	-	166 µg/100g of edible protein
Vitamin C	-	5mg/100gm

❑❑❑

Chapter 9

Nutritional Quality of Vegetables I

Tomato

It is known for its outstanding nutritive value as indicated below.

Composition of tomato fruit per 100 g of edible protein.

Moisture	-	93.1g
Protein	-	1.9g
Fat	-	0.1g
Minerals	-	0.6g
Fibre	-	0.7g
Carbohydrates	-	3.6g
Sodium	-	45.8mg
Potassium	-	114mg
Copper	-	0.19mg
Sulphur	-	24mg
Chlorine	-	38mg
Vitamin A	-	320 I.U.
Thiamine	-	0.07mg
Riboflavin	-	0.01mg
Nicotinic acid	-	0.4mg
Vitamin C	-	31mg
Calcium	-	20mg
Magnesium	-	15mg
Oxalic acid	-	2mg
Phosphorus	-	36mg
Iron	-	1.8mg
Acidity	-	4.2-10.2mg/100 ml
Ascorbic acid	-	20.9-22.5mg/100g

It is also rich in medicinal value. The pulp and juice are digestible, a promoter of gastric secretion and blood purifier. It is also considered to be intestinal antiseptic. Dried tomato juice retains vitamin C. Tomatine content in small hard tomato is 300mg/100g of edible protein and in larger green fruits the content is 150mg/100g. Light cooking decreases the alkaloid slightly but four weeks storage of freeze-dried product reduces 5 to 10 per cent of the original level. The solanene content of the pulp made from unripe tomato is much lower (5mg/100g).

In the process of ripening, the green chlorophyll pigment is degraded and yellow orange carotenoid and red lycopene are synthesized. The biosynthesis of these pigments are light and temperature dependent. Carotene and lycopene development are promoted by red light and inhibited by far red, indicating that the process is under phytochrome control.

The acid content was found to be lower in immature fruit and it was the highest at the stages when colour appeared with a rapid decrease when the fruit ripened. Citric acid was the major constituent of total acid. The total soluble solids and reducing sugars increased throughout the development of fruit. Both β-carotene and lycopene increased in concentration as the fruits developed from mature green into ripening stages.

It is excellent source of Vitamin C, so it is known as poor man's orange. Popular vegetable salad, tomato juice is a popular appetizer and beverage.

Seed contains 24% oil. The semidrying oil is used as salad oil and in the manufacture of margarine.

Okra (bhendi)

Carbohydrates	-	14.5%
Mineral matter	-	0.7%
Calcium	-	10mg/100g
Mg	-	12mg/100g
Phosphorus	-	70mg/100g
Iron	-	0.3mg/100g
Oxalic acid	-	14mg/100g
Thiamine	-	0.06mg/100g
Riboflavin	-	0.1mg/100g
Nicotinic acid	-	0.3mg/100g
Vitamin C	-	14mg/100g

Its tender green fruits are used as a vegetable :

Moisture	-	89.6g
Protein	-	1.9g
Fat	-	0.2g
Minerals	-	0.7g
Fibre	-	1.2g
Carbohydrates	-	6.4g
Sodium	-	7mg
Potassium	-	103mg
Copper	-	0.19mg
Sulphur	-	30mg
Phosphorus	-	56mg
Iron	-	1.5mg
Calcium	-	66mg
Magnesium	-	43mg
Vitamin A	-	88 I.U.
Thiamine	-	0.07mg
Riboflavin	-	0.10mg
Nicotinic acid	-	0.60mg
Vitamin C	-	13mg

It is rich in vitamins, Ca, K and other mineral matters. The roots and stems of okra are used for clarifying the cane juice from which gur or brown sugar is prepared. Its ripe seeds are roasted ground and used as a substitute for coffee in Turkey. Matured fruits and stem containing crude fibre are used in the paper industry.

Generally, the fibre formation in the fruit starts from the 6th day of fruit formation and a sudden increase in fibre content from the 9th day is observed.

Brinjal (egg plant)

It has been a staple vegetable in our diet since ancient times. It is quite high in nutritive value.

Moisture	-	92.7g
Protein	-	1.4g
Fat	-	0.3g

Minerals	-	0.3g
Fibre	-	1.3g
Carbohydrates	-	4.0g
Sodium	-	3mg
Potassium	-	2mg
Copper	-	0.17mg
Sulphur	-	44mg
Phosphorus	-	47mg
Iron	-	0.9mg
Chlorine	-	52mg
Calcium	-	18mg
Magnesium	-	16mg
Vitamin A	-	124 I.U.
Thiamine	-	0.04mg
Riboflavin	-	0.11mg
Nicotinic acid	-	0.09mg
Vitamin C	-	12mg

The oblong fruited brinjal cultivars are rich in total and water soluble sugars, whereas the long fruited cultivars contain large amounts of reducing sugars, anthocyanin, phenols, glycoalkaloids (as solasodine), dry matter and amide protein. A high anthocyanin content and a low glycoalkaloid content are considered essential, regardless of how the fruit is to be used. For processing purposes, the fruit should have a high dry matter content and a low level of phenolics. On an average the round types of brinjal had higher polyphenol oxidase activity and glycoalkaloid content than long types. Bitterness in brinjal is due to presence of glycoalkaloids. Generally high amounts of glycoalkaloids (20 mg.100g fresh wt) produces a bitter taste flavour. Usually glycoalkaloid contents vary from 0.37 mg per 100g fresh weight to 4.83 mg in most of the commercial cultivars of India. The discolouration in brinjal fruits is attributed to high polyphenol oxidase activity. Copper content and polyphenol oxidase activity were highest in the green and lowest in white cultivar. Green cultivar had better processing properties than the more popular purple cultivar. White cultivar contained twice as much as crude fibre as the purple and green cultivars.

Nutritive value of different gourds

Name	Pumpkin	Snake gourd	Ridge gourd	Cucumber	Chow chow	Bottle gourd	Bitter gourd	Ash gourd
Moisture %	92.6	94.6	95.2	96.3	92.5	96.1	92.4	96.5
Protein %	1.4	0.5	0.5	0.4	0.7	0.2	1.6	0.4
Fat %	0.1	0.3	0.1	0.1	0.1	0.1	0.2	0.1
Minerals %	0.6	0.5	0.5	0.3	0.4	0.5	0.8	0.3
Fibre %	0.7	0.8	0.5	0.4	0.6	0.6	0.8	0.8
Carbohydrates %	4.6	3.3	3.4	2.5	5.7	2.5	4.2	1.9
Ca (mg)	10	26	18	10	140	20	20	30
Phosphorus (mg)	30	20	26	25	30	10	70	20
Iron (mg)	0.7	0.3	0.5	105	0.6	0.7	1.8	0.8
Carotene μg/100g	50	96	33	-	-	-	126	-
Thiamine mg	0.06	0.04	-	0.03	-	0.3	0.07	0.06
Riboflavin mg	0.04	0.06	0.01	-	0.04	0.01	0.09	0.01
Niacin mg	0.5	0.3	0.2	0.2	0.4	0.2	0.5	0.4
Vitamin mg	2	-	5	7	4	-	88	1

Bitter gourd is rich in vitamin C, pumpkin containing high carotenoid pigments and chow-chow fairly high in calcium.

Cucurbitacins

Biochemically the cucurbits are characterized by bitter principles, called cucurbitacins. They occur in nature, free as glycosides. Highest concentrations are found in fruits and roots, leaves being normally slightlybitter. Bitter principles found in roots differ from those in fruits. Seedlings, especially radicles contain Cucurbitacins which are considered primary and hence significant. In a plant three situations can occur:

(i) Fruit and vegetative parts bitter
(ii) Fruit not bitter but vegetative parts bitter and
(iii) Bitterness absent in both vegetative parts and fruits.

The nutritive value of seeds is due to their high oil and protein content. The seed oil of cucurbitaceae falls into two groups: one with the palmitic-oleic-linoleic acid composition, the other with conjugated triene acids, purnicic acid, alpha-oleostearic acid seed proteins of cucurbitaceae are comparable in nutritive value to those of leguminosae, though like other plant proteins they tend to be low in lysine and sulphur containing amino acids. They are generally rich in methionine. Atleast 150 fatty acids are known to occur as components of seed oils. Genera like cucumin (cucumber) and cucurbita have usually ordinary unsaturated acids (oleic and linoleic) while others like momordica and trichosanthes are noted for acids with conjugated double bonds (pumicic and alpha-oleostearic).

Amaranthus

The leaves and tender stems of amaranthus are rich in protein, minerals, vitamin A and C. The composition is as follows :

Moisture	-	85.7%
Protein	-	4.0%
Fat	-	0.5%
Minerals	-	2.7%
Fibre	-	1.0%
Carbohydrates	-	6.3%
Sodium	-	230mg

Potassium	-	341mg
Sulphur	-	61mg
Phosphorus	-	83mg
Iron	-	25.5mg
Magnesium	-	247mg
Vitamin A	-	9200 I.U.
Thiamine	-	0.03mg
Riboflavin	-	0.10mg
Nicotinic acid	-	1mg
Vitamin C	-	99mg
Calcium	-	3.97mg

The fresh tender leaves and stem of amaranthus are delicious when cooked like other fresh leafy vegetables.

Drumstick

	Drumstick	Drumstick leaves
Moisture	86.9%	75.9%
Carbohydrates	3.7%	12.5%
Protein	2./5%	6.7%
Fat	0.1%	1.7%
β carotene	110μg	6780μg
Thiamine	0.05mg	0.06mg
Riboflavin	0.07mg	0.05mg
Vitamin C	120mg	220mg
Calcium	30mg	440mg
Iron	5.3mg	7mg

❑❑❑

Chapter 10

Nutritional Quality of Vegetables II

Cabbage

Moisture	-	92.4%
Protein	-	1.4%
Fat	-	0.5%
Minerals	-	2.7%
Fibre	-	1.0%
Carbohydrates	-	5.3%
Potassium	-	2.71%
Sulphur	-	1.16%
Phosphorus	-	2.71%
Iron	-	205ppm
Copper	-	15ppm
Iodine	-	218ppm
Calcium	-	0.73%
Vitamin A	-	400 I.U.
Vitamin B_1	-	27.1 I.U.
Vitamin C	-	100mg

Cauliflower

Moisture	-	91.7%
Protein	-	2.4%
Carbohydrates	-	4.9%
Fat	-	0.2%
Calcium	-	0.35%
Phosphorus	-	0.76%
Potassium	-	3.58%

Sulphur	-	1.07%
Iron	-	117ppm
Copper	-	36ppm
Vitamin A	-	70 I.U.
Vitamin B_1	-	56 I.U.
Vitamin C	-	75 mg

Cole crops are known for its rich source of vitamins A and C. It also contains minerals including phosphorus, potassium, Ca, Na and Fe.

Radish

Nutritional value of different cultivars and different parts of radish

Parameters	Radish roots (%)		Radish Tops (%)
	White cultivar	Pink cultivar	
Moisture	94.4	90.8	90.3
Protein	0.7	0.6	2.7
Fat	0.1	0.3	0.6
Fibre	0.8	0.6	0.9
Carbohydrates	3.4	6.8	3.4
Minerals	0.6	0.9	2.1
Calcium	50mg	50mg	310mg
Phosphorus	22mg	20mg	60mg
Iron	0.4mg	0.5mg	16.1mg
Vitamin A	5 I.U.	5 I.U.	18.6 I.U.
Thiamine	0.06mg	0.06mg	0.03mg
Riboflav in	0.02mg	0.02mg	0.16mg
Nicotinic acid	0.5mg	0.4mg	0.3mg
Vitamin C	15mg	17mg	103mg

It is a good source of vitamin C, containing 15 to 40 mg/100g of edible portion and supplies a variety of minerals. Pink-skinned radish is generally richer in ascorbic acid than the white-skinned one. Radish contains glucose as the major sugar and smaller quantities of fructose and sucrose. Pectin and pentosans are also reported to be present. The characteristic pungent flavour of radish is due to the presence of volatile isothiocyanates [trans-4-methyl-thiobutenyl] and the colour of the pink cultivars is due to the presence of anthocyanin pigments.

The leaves of radish are a good source for extraction of proteins on a commercial scale and the radish seeds are a potential source of a non-drying fatty oil suitable for soap making, illuminating and edible purposes.

Peas

It is a very common nutritious vegetable grown in the cool season throughout the world.

Moisture	-	72%
Protein	-	7.2%
Fat	-	0.1%
Fibre	-	4%
Carbohydrates	-	15.8%
Minerals	-	0.8%

Different Minerals Compositions mg/100 g

Magnesium	-	34mg
Phosphorus	-	130mg
Sodium	-	7.8mg
Calcium	-	20mg
Potassium	-	79mg
Sulphur	-	95mg
Iron	-	1.5mg
Copper	-	0.23mg
Vitamin A	-	139 I.U.
Thiamine	-	0.25mg
Riboflavin	-	0.01mg

Nicotinic acid	-	0.8mg
Vitamin C	-	9 mg
Oxalic acid	-	14mg

It is very rich in protein, carbohydrate vitamin A and C, calcium and Phosphorus. Large proportion of peas are processed (cannel, frozen or dehydrated) for consumption in the off season. Pea-straw is a nutritious fodder. It is very rich in protein and, therefore, highly valuable for the vegetarians.

Potato

Moisture	-	74.7%
Protein	-	1.6%
Carbohydrates	-	22.6%
Fat	-	0.1%
β carotene	-	24 μg
Riboflavin	-	0.1mg
Thiamine	-	0.1mg
Vitamin C	-	17mg
Calcium	-	10mg
Iron	-	0.7mg

It is rich in carbohydrates and used as a supplement for energy source.

❑❑❑

Chapter 11

Chemical Constituent of Spices and Condiment

They are the green or dried ripe fruits of pungent forms of *Capsicum annuum* and sometimes *Capsicum frutescens.* Pungency together with colour determine the quality of product. Major compound responsible for the characteristic aroma of the fruit is 2 methoxy 3-isobutyl pyrasine.

Composition (Green chilli)

Moisture	-	85.7%
Protein	-	2.9%
Carbohydrates	-	3.0%
Fat	-	0.6%
Minerals	-	1.0%
Fibre	-	6.8%
Calcium	-	30mg
Magnesium	-	24mg
Potassium	-	217mgf
Phosphorus	-	80mg
Sodium	-	6.5mg
Sulphur	-	34mg
Chlorine	-	15mg
Iron	-	1.2mg
Riboflavin	-	0.39mg
Nicotinic acid	-	0.9mg
Thiamine	-	0.19mg
Uitamin A	-	292 I.U.
Vitamin C	-	111mg

The main constituent of chilli pericarp is a crystalline, colourless pungent principle known as capsaicin or capsicutin ($C_{18} H_{27} NO_3$) a condensation product of 3-hydroxy-4-methoxy benzylamine and decylenic acid which produces a highly irritating vapour on heating. It is secreted by the outer walls of the fruit. The content of capsaicin differs in cultivars of chillies. Green chillies are rich in vitamin A and C and the seed contains traces of starch. Capsaicin content of red chilli is 0.34% (dry weight) conical thinner pericarp type has greater pungency than rounder thicker pericarp. Capsaicin level increases as fruit matures and red colour is due to carotenoid pigment of which most important one is capsanthin ($C_{40} H_{58} O_3$)-keto carotenoid pigment. Another contributor is capsorubin.

Garlic Composition

Moisture	-	62.8%
Protein	-	6.3%
Fat	-	0.1%
Mineral matter	-	1.0%
Fibre	-	0.8%
Carbohydrates	-	29.0%
Calcium	-	0.03%
Phosphorus	-	0.31%
Iron	-	0.001%

It has been considered as rich source of carbohydrates, protein and phosphorus. Ascorbic acid content was reported to be very high in green garlic. The uninjured bulk contains a colourless, odourless water soluble amino acid allinin. On crushing the garlic bulb, the enzyme allinase breaks down allinin to produce allicin of which the principal ingredient is the odoriferous diallyldisulfide. It also contained about 0.1% volatile oil. The main constituent of the oil are; diallyl disulfide (60%), diallyl trisulfide (20%), allyl-propyl disulfide (6%) a small quantity of diethyl disulfide and probably diallyl polysulfide. Diallyl disulfide is said to possess the true garlic odour. An antibiotic allicin is the active principle in garlic which is an enzymatic cleavage product from its precursors allinin, naturally, present in garlic.

Spices

Plant products used as food adjuncts to add aroma and flavour. Eg. Pepper, cardamom.

Condiments

Adjuncts plant products used as food adjusts to add taste. Eg. Chillies, coriander etc.

Both species and condiments contain essential oil which provide flavour and taste. They are of little nutritive value.

Pepper : *Piper nigrum*

Two products

Internationally traded

Black pepper	White pepper
↓	↓
Prepared from dried immature berries	Prepared by removal of mesocarp of matured berries

Black pepper

Composition

Moisture	-	18.7 to 21.7%
Total N	-	1.55 to 2.67%
Starch	-	28 to 49%
Crude fibre	-	8.7 to 18%
Crude piperine	-	4 to 10%
Volatile oil	-	2 to 3.8%
Ash	-	2.6 to 5.6%

Black pepper constitute 34% starch and white pepper has 56% starch but less of fibre.

Quality is assessed on content of aromatic steam volatile oil and of pungent alkaloid (expressed as piperine). A considerable portion of N consists of non-proteinaceous form such as alkaloid. Half of protein is made up of simple amino acid. The alkaloid piperine is responsible for the pungent taste of black and white pepper. It is hydrolysed into peperidine and piperic quantity are peperline, piperettine, piperolin A and B, piperanine. Piperettine on hydrolysis yields piperidine and isopiperttine. The characteristic aromatic odour of pepper is due to volatile oil in cells of pericarp.

Black pepper - 0.6 to 2.6%

White pepper - 0.5 to 1.0%

This volatile oil composed mainly of monoterpene hydrocarbon together with smaller amount of sesquiterpene hydrocarbon.

Piperine formula is C_{17} H_{19} O_3 N.

Coriander

Plant yield two products used for flavouring purpose: 1. Fresh green leaves. 2. Riped fruit. Odour and flavour of the two products are markedly different. Leaves are used for soup making and fruits are used for the preparation of curry powder. Major constituents of dried seed are carbohydrate, fibre, fatty oil and protein. The quality determinate of spice is content and composition of steam volatile oil.

In unriped fruit and vegetative parts of plants aliphatic aldehydes dominate the steam volatile oil and this is responsible for flavour and odour. On ripening fruits, acquire pleasant and sweet odour. Major constituent is monoterpene. Alcohol present is linolol. The content varies from 0.1 to 1.7%. Fruits contain more volatile oil canals. The odour, composition and content of volatile oil change during the course of ripening. In unripe fruits two types of volatile oil canals are present. One type is located as periphery of canal while contains volatile oil similar to that present in the vegetative organ. Berries in the mesocarp of fruit contains linalool as important constituent. As fruit ripens the peripheral canals flattened, commence to lose volatile oil and odour of fruit changes. On drying to 7% moisture, the outer canal completely lose volatile oil but the inner canal remain intact and the characteristic odour composition of the volatile oil of the fruit is attained. Fresh leaves contain higher amount of vitamin C and carotene. Vitamin C decreases on storage. Seeds contain a higher amount of vitamin A.

Ginger

It is used in medicine as carminative and aromatic stimulant to gastro intestinal tract. Ginger rhizome contains little steam volatile oil, pungent compound, resin, protein, cellulose, pentasan, starch and mineral elements. Fibre, volatile oil content and pungency level are most important criteria in assessing the suitability of ginger for processing purpose. Relative abundance of the above three compounds in fresh rhizome is governed by state of maturity at harvest. Crude fibre content of ginger is 10%. Pungent principle is 1 to 2% in freshly prepared dried ginger. The pungent principle is due to the presence of a resin - a brown lignin contains the active principle ginevol. Minor ones present are zingerone, shognol, etc. Aroma and flavour are determined by the composition of steam volatile oil comprised mainly of sesquiterpene hydrocarbon, monoterpene hydrocarbon and oxygenated monoterpenes zingiberene-sesquiterpene hydrocarbon zingiberol-sesquiterpene alcohol. These two are responsible for aroma.

❑❑❑

Chapter 12

Chemical Constituent of Narcotics

Chemistry of Narcotics

Tobacco

Main ingredient which gives distinct property is nicotine, pyridine group of alkaloid. Total alkaloid content varies from 4 to 6% and never exceed 6%. Content varies with different manufactured type.

Pipe tobacco	-	0.5 to 0.8%
Cigar	-	0.8 to 2.09%
Cigarette	-	3%
Chewing type	-	3 to 6%

Nicotine is present in combination with malic and citric acids. It is a virulent poison acting mainly on nervous tissue, nerve centre and paralyzing them.

Nicotine ($C_{10}H_{14}N_2$)

There are isomers of nicotine. They are 1. Nornicotine, 2. Neonicotine or Anabasine.

Nicotine occurs in *Nicotina tobaccum* and *Nicotina rustica*. It is extracted from stem and midrib of tobacco plant using benzene as extractant. After the extraction, the extract is treated with lime and caustic soda to release the alkaloid and then treated with H_2SO_4 and converted to nicotine sulphate. Subsequently crystallized and used for the manufacture of insecticides.

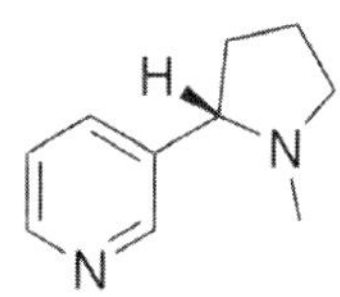

Properties of nicotine

- In pure condition it is colourless, oily liquid
- In pure form it is soluble in water and organic solvents

- It is volatile in nature and so has a fumigant action
- When exposed to air, turns to brown material due to antioxidation
- It has a burning taste and basic in reaction.

Nornicotine ($C_9 H_{12} N_2$)

Removal of methyl group by hydrogen.

It occurs in *Nicotina silvestris*. It is similar to nicotine except that methyl group in pyridine ring is replaced by H atoms. Physical and chemical properties are similar to nicotine except it is less volatile and more stable.

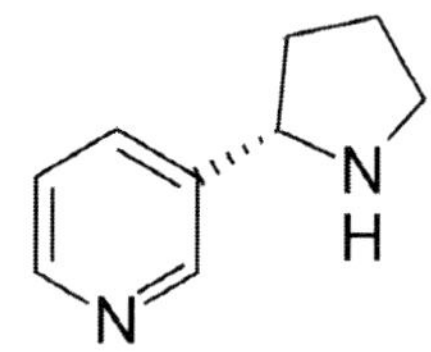

Neonicotine ($C_9 H_{10} N_2$)

It occurs in *Nicotina glaucum* and *Nicotina anabasin*. It resembles the other two in physical and chemical properties. It is less volatile and fairly stable than the other two.

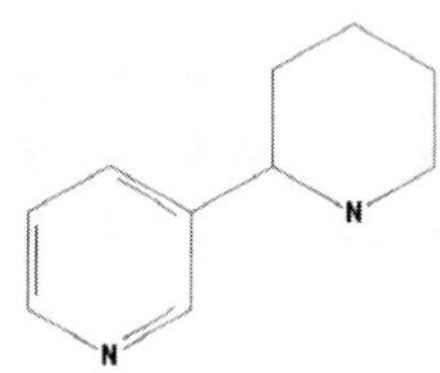

Chemical composition

Freshly harvested leaves contain 85 to 90% moisture which is reduced to 80 to 85% on wilting and to 10 to 25% on complete curing.

Carbohydrate

It comprise 25 to 50% of total drymatter of mature leaf of which starch and invert sugar is important constituent and relative increase in cure tobacco and decrease in cigar type.

Nitrogenous compounds

In green leaves most of the N_2 is protein fraction with nicotin.

Polyphenols

Tannin like substance occur in leaves and affect cell in air and on fire curing. They are oxidized giving red and brown colour to the leaves.

Chlorophyll and other pigments

During early phases of curing green pigments chlorophyll- A and B are destroyed by oxidation and yellow pigments carotene and xanthophylls are present.

Mineral constituent

The ash content varies between 12 to 25% on dry weight basis. Quantity and composition of mineral component affect the combustability of leaves.

□□□

Chapter 13

Chemical Constituent of Beverages

Chemistry of beverages

Coffee

The important constituent which acts as stimulant in body is caffeine, the alkaloid present in tea and cocoa to smaller extent. Leaves and beans of coffee contains caffeine (caffeine is 1,3,7 trimethyl xanthino). Formula is $C_9H_{10}N_4O_2$.

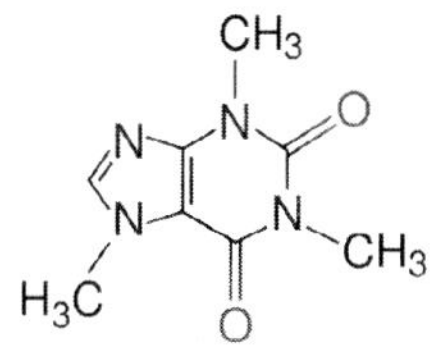

Percentage of caffeine content varies with species and varieties *Coffea arabica* and *Coffea liberica* contains 1.2 to 1.8% caffeine.

Coffea robusta contains 2 to 2.6% caffeine . Some species is free from caffeine content (Madagascar coffee). Caffeine undergoes little loss only during roasting. Roasting brings about rupture of oil canal in bean resulting in escape of oil as evidence by oily coat on roasted bean. There is considerable formation of gases especially CO_2, a product of oxidation, part of which is held in the bean itself giving it slightly a puffed up appearance as compared with raw bean. Total oil content in the raw bean is 12 to 13%, of this the volatile oil escape during roasting. Aroma and flavour associated with roasted bean and it is mainly due to the presence of essential oil called caffiol which is water soluble on infusion.

Chemical Composition of Coffee

	Constituent	Raw bean	Roasted bean
(i)	Moisture	8.75%	3.75%
(ii)	Ash	4.41%	4.49%
(iii)	Oil	12.96%	13.76%
(iv)	Caffeine	1.87%	1.81%
(v)	Crude fibre	20.7%	14.7%
(vi)	Protein	9.5%	12.9%

Tea

Most important constituents which give tea as a beverage are polyphenol, caffeine and essential oil (caffeine is also called theine). Fresh packed tea contains 75 to 80% water, soluble and insoluble constituents-25%, protein 20%, crude fibre-27%, carbohydrate-4%, Ash content-4.5 to 5%, pectin-6%, caffeine-2.5 to 4.5. Cell sap contains higher amount of K salt.

The most important and characteristic component in the cell sap is polyphenol which undergoes a series of changes when the leaf is macerated during the manufacture of tea-cured.

- Black Tea
- Green Tea

Those occurring in tea are gallic acid and catechin which are oxidized by enzyme during maceration and fermentation to produce orthoquinenes which polymerise to produce colour astringent condensation product and these are partially extracted tea. The four types of catechin which occur naturally in tea leaves are

1. Catechin,
2. Epicatechin,
3. Galocatechin and
4. Epigalocatechin.

Other phenolic compounds present are ratin, chlorgenic acid and theogallia. They are all the corresponding ester of quinic acid. The polyphenol and caffeine content highest in the buds and top leaf and successively diminished in lower leaf and stock. Good quality tea is a product of good leaf with higher polyphenol and increased enzyme activity. Tea grown under shade has decreased the polyphenol content.

Pigments

Chlorophyll of leaves enter into chemical reaction and decompose during process of fermentation. The red and yellow pigments of tea leaves are derived from anthocyanins and flavones. Green tea is made from tea leaves which are steamed and dried and does not undergo withering and fermentation. Black tea is made from withered, rolled, fermented and dried tea leaves. In black tea manufacture, polyphenol (tannins) are oxidized by enzymes in the presence of air, the bruised condition of old leaves and all sap is brought to the surface. Tannin is oxidized and reduced in quantity. Manufactured tea contains only 4.5% tannin as against 13 to 14% tannins present in fresh leaves.

Green tea avoids the above procedure and tannin is not lost to some extent. Too long soaking of tea leaves in hot water dissolve greater proportion of tannin and make tea bitter in taste. All beverage crops contain xantheine compound which is responsible for imparting particular flavour. By way of processing we can get maximum flavour. Processing can be done by 1. Fermentation, 2. Heating/roasting/ thermal treatment.

Fermentation

Tissues of tea leaves or coffee beans is subject to enzymatic degradation with ethanol.

Chemical changes during fermentation

Most important flavour producing agent is trigonelin (3-carboxyl-ethyl pyridimin hydroxide and some compound like glucose, fructose, amino acids and pectin). Due to oxidative fermentation, flavoids like epicatechin and epigalocatechin are converted to thioflavin. Oxidative degradation of leucine, isoleucine and valine takes place.

Thermal treatment

Firing - Tea

Roasting - Coffee

During roasting pyridine and pyroïd compounds are produced. Citric and malic acids are partly degraded into aconitic, fumaric and citrogenic acids.

❑❑❑

Chapter 14

Chemical Constituent of Essential Oils

Essential oil

These substances with characteristic odour and resinous property are extracted from special type of glandular hair/cells present in leaves, stem, bark etc. They can be steam distilled and produce temporary greasy spot on paper.

Eucalyptus

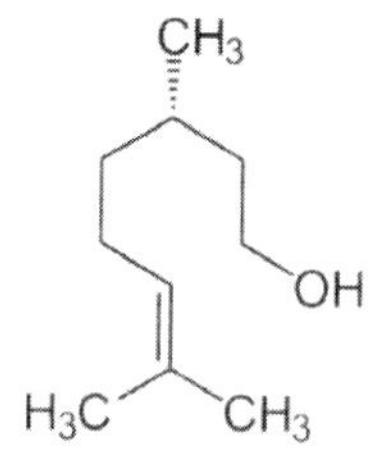

They are native of Australia. The term Eucalyptus oil denotes three distianct group of essential oils viz., medicinal type, perfumery type and phellandrence-rich type blue gum (*Eucalyptus globules*) belongs to the first group which has high cineole content and has wide medicinal application. *Eucalyptus citriodora* belong to the second and third groups which are characterized by their high citronellal and phellandrene contents respectively. The oil is widely used in soap, perfumery and cosmetic industries and for the isolation of hydroxyl citronellal used in the manufacture of high grade perfumes. The leaves yield 1% oil on an average. The citronellal and cineole contents of oil varies from 75 to 85%.

Cinchona

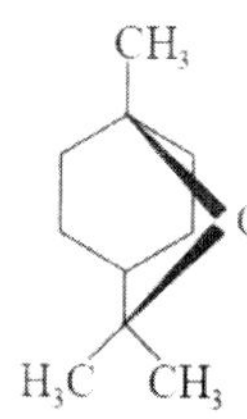

Cineole

Quinidine an alkaloid of cinchona bark is used for its anti-malarial action. In addition to its use in pharmacy, quinine and quinidine and their derivatives are utilized in insecticide formulations for the preservation of fur, feather, wool and textiles.

Geranium

The oil of geranium is widely used in many perfumery and cosmetic industries due to its agreeable and very profound strong rose-like odour. If pure, the oil is almost a perfume by itself and blends well into all kinds of scents. It is widely used in scenting of soaps. Tannin can also be obtained as a by-product from geranium stems and leaves after extraction of oil.

The freshly harvested terminals are used for distillation of oil. The oil content of geranium ranges from 0.08 to 0.15 per cent and is much during summer months from April to June. High humidity and rains at the time of harvest lowers the yield of oil. The terminal portion with 6 to 12 leaves contain more oil than the middle and basal portions.

The geranium oil posses a strong rose-like odour. The main constituents of the oil are geraniol (68 to 75%) and citronellel (23-40%). The quantity of geraniol and citronellol their ratio vary in different geranium oils. Esters of these alcohols as formate and tigilate also occurs.

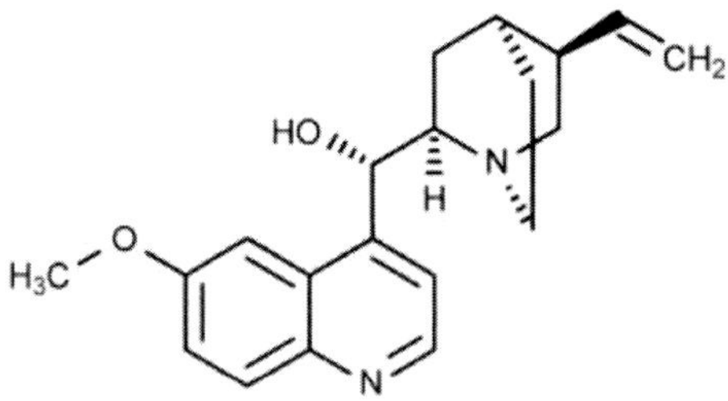

Properties of Geranium oil of different places of Tamil Nadu

S.No.	Properties	Yercaud	Nilgiris
1.	Specific gravity (at 15 °C)	0.899	0.894
2.	Optical rotation	-	7.5°
3.	Refractive index	1.464	1.471
4.	Esters %	19 to 22.3	23
5.	Total alcohols as geraniols (%)	-	66.4
6.	Solubility 70% alcohol	1:3 vol	1:3 vol

❑❑❑

Chapter 15

Nutrional Quality of Forage Crops

Chemistry of forage crops

The plants which are cultivated to feed domestic animals are forage crops. There are four categories:

- Cereals
- Legumes
- Grasses
- Tree

Cereals and grasses

They are high in carbohydrate and less in protein and minerals. They contain 6 to 8% protein and 30% crude fibre.

Legumes

They are high in protein and minerals but less in carbohydrate. Crude fibre content is around 20%.

Suitability of forage crops

- It should contain greater total digestible nutrient (TDN)
- It should give more yield
- It should be cultivated in all seasons
- Crops should contain greater amount of nutrients and free from toxic elements.
- Crop should have digestible nutrients

Normally forage comprises of grasses, hay, straw, silages, tubers, roots, etc. Grasses and legumes are important forage crops. Legumes contain higher protein and calcium contents and lower fibre content than grasses at comparable stages of growth.

Chemical composition

Nitrogenous substances

NO_3 content of forage crop is less than 1% of dry weight. Among nitrogenous substances proteins, amino acids, amides, nitrogenous bases etc. occur.

Hydrocyanic acid

It is a nitrogen containing glucoside which liberates HCN on hydrolysis. "Dhurrin" a glucoside, is found in young seedlings of cholam and sudan grass. An enzyme, Dhurrinase acts on this glucoside and releases HCN. The HCN content in different plant parts vary from 6 to 350 mg/100g of dry weight. Internal factors controlling quantity of HCN are heritable and reduction of HCN is effected by suitable breeding.

Protein

It occurs abundantly in leaves and other parts of plants. Amino acids like histidine, leucine, lysine, tryptophan and arginine are also present. Chlorophyll content is correlated with protein content and carotene content.

Nitrogen free extract (CHO)

Total sugars are greater in young grass. Water soluble carbohydrates are more in grasses than in legumes which make them highly suitable for silage. Glucosyl and tannins are also occur. Crude fibre of forage consists of 85% cellulose and 12% lignin. Lignin content usually varies from 2 to 20%. On the cellulose frame work, hemicellulose and lignin are deposited.

Pectins

It is a major constituents in grass amounting to < 1% and is located in the middle lamellae of cell wall.

Lipids / ether extractive (EE)

About 35% of crude fat is saponifiable matter in hays and fodders. As grass mature both crude fat and total fatty acids decrease. The lipids comprise a fairly complex mixture of fats, waxes, and phospholipids, glycolipids and sterols, all of which with the exception of sterols, contain fatty acids in their chemical structure. Grass herbage usually contains 4 to 8% of lipid material. Fertilizer N at high rates increases fatty acids and sterols. Legumes appear to have similar lipid contents as in grasses.

The main organic acid present in grass herbage are malic, citric, quinic and succinic. NO_3 application increases the organic acid contents. Clovers and lucerne appear to contain greater quantities of organic acids than grasses.

Carotene occurs in considerable quantity (20 to 40 mg/kg) in the dry matter. It decreases with maturity. Vitamin B complex is higher in legumes than in grasses.

Vitamin D is present in sun-cured hays. Vitamin E content falls markedly with increasing maturity from 20 mg/100g to 1.5 mg/100g. It is mostly present in leaves. Oestrogenic compounds are mostly present in legumes. They influence the reproductive behaviours of animals.

In goistrogenic compounds, cyanogenic glucosides appear to be the main source of goistrogenic activity. They are hydrolysed in the rumen with the release of hydrocynic acid and this inturn is converted into thiocyanate, which inhibits the uptake of iodine by the thyroid gland.

Mineral elements such as P, K, Ca, Mg, S, Na, Cl, Fe, Mn, Cu, Zn and Co are present in forage crops. Mo at high level may interfere with Cu metabolism. Mg and Cu contents are high in legumes than in grasses. Fe content is more in legumes than in grasses whereas Mn content is lower in legumes than in grasses generally contain more Co than grasses growing under similar conditions.

In general legumes contain lesser cell wall material and dry matter content than the non-legumes at the same stage of maturity. The cell walls of legumes have considerably less hemicellulose content, but are more lignified. Non-legumes have more cellulose and silica.

Quality evaluation of forage crops

The output per animal is a function of the nutritive value of forage and the rate of its consumption. Hence "the output per animal is related to the total character of the forage and it is in this context the term forage quality is used". The rate of consumption of digestible dry matter has come into common usage these days to express forage quality.

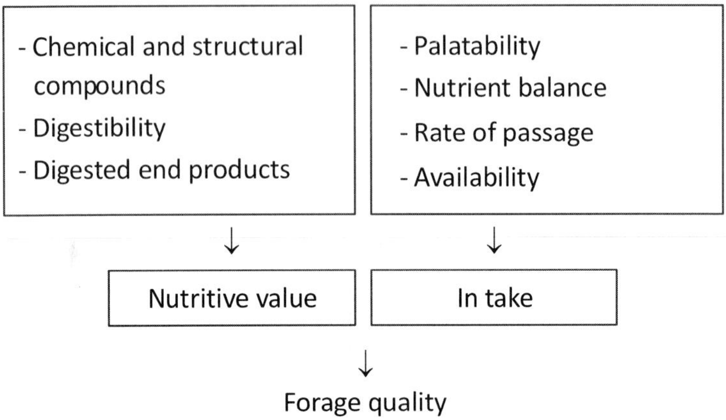

This again can be very suitably explained through a diagram.

Forage that is too fibrous for human, poultry, pigs, is eaten by cattle, buffaloes, camels, goats, sheep, etc. The term forage covers a wide range of feeds, natural grasses, sown grasses and legumes, tree leaves, sorghum, maize, berseem and a wide range of by-products including straws, corn stovers, pea haulms, etc., the common feature is their relatively high level of fibre, which is partially digested by the symbiotic bacteria living in the enlarged fore stomach of the ruminant. The fibre content in forages varies widely according to species, varieties, stages of maturity and the environmental conditions.

Voluntary intake is defined as the amount eaten during a period of time when the forage is offered and is expressed as g/day. It is governed by nutritive value of the forage pasture, contrasting leaf and stem, amount of legumes in the pasture, confinement of the animals to graze only in restricted areas, etc.

Forage quality or feeding value could be defined as a biological assessment of the work of a forage in terms of animal production. Forage quality = Animal production + (intake x utilization). Since the product of energy intake (I) and the utilization (U) is also the amount of energy available for maintenance (M) and production (P). Nutrient intake = intake of feed x digestibility of feed x efficiency of utilization of digested feed.

The two systems scientifically advocated these days for feed evaluation are based on NE (Net Energy) and ME (Metabolizable Energy). Digestible energy is the gross heat of combustion of the food minus the gross heat of combustion of the faeces. Energy lost in urine and in combustible gases are subtracted from the digestible energy (DE), gives ME. Multiplying DE by a factor 0.82, we can calculate ME. NE is that proportion of GE of feed that is used by the animal for maintenance and production.

% Digestible Energy = 1.09% DMD-8.9

(DMD = dry matter digestibility)

D-value is the percentage of digestible organic matter

And it is = 0.926%

DMD-0.403% ash + 4.6372

The term digestibility refers to the extent to which any feed or feed component is digested by an animal. The quantity digested is termed as the digestibility coefficient.

Parallel to the development of chemical methods of forage evaluation, there has been the development of biological methods of evaluation, the artificial rumen or in vitro digestion method. This is based on incubation under controlled conditions, of a sample of the test forage with a mixed culture of the micro-flora taken from the rumen of a forage-fed animal. In an alternative system, a sample of dried forage is enclosed in a nylon or Dacron mesh bag suspended within the rumen in vivo, and digestibility and rate of digestion are measured by the loss of dry matter.

Sorghum

Cyanogenic glycosides called Dhurrin and Linnamarin are present in sorghum. During mastigation the enzyme linnamarese combine with dhurrin and produce HCN. HCN prevent transport of O_2 in animal system and leads to the death of animals. The threshold content of HCN in sorghum is 100 ppm.

Hybrid-Napier grass

Application of fertilizers to napier grass leads to production of high amount of Na and NH_4 oxalate in the plant. This is produced because of imbalance between anions and cations in the plant. If plants containing high amount of Na and NH_4 oxalate are taken by animals, oxalic acid is produced in the animal system. In the stomach, oxalic acid combines with Ca and results in the formation of Ca-oxalate which aids in the stone formation in the kidneys and also results in the formation of harmful substances in animals. Threshold value of oxalic acid is 3%.

Sorghum

Herbage quality depends mainly on the stage of growth at which the crop is cut and partly on the amount of fertilizer N applied. The percentage of crude protein was the highest at boot leaf stage and also the carbohydrate value.

The chemical composition of sorghum herbage on dry matter basis as follows.

Crude Protein (CP)	- 7.75%
Ether extractives (EE)	- 1.73%
Crude fibre (CF)	- 32.36%
Total ash	- 8.55%
Nitrogen free extract (NFE)	- 49.61%

Crude protein digestibility in silage is lower than that in green fodder. In dried or ensoled sorghum the content of HCN decreased sharply or hay presents no danger to animals.

Appreciable quantities of cyanogenic of glucosides , called Dhurrin occur in sorghum species. On enzymic hydrolysis, Dhurrin yields aghycone (methyl glucosinolates) which spontaneously dissociates to hydrogen cyanide and aldehyde. These compounds are toxic and affect palatability of the fodder. In young herbage, the HCN content may be as high as 750 ppm or more which is dangerous to animals. The crop should preferably be harvested after flowering but never before 40 to 50 days from the date of sowing. Limited supply of N or condition it increases. The leaves contain higher amounts of glucosides than the stems. On dry weight basis 500 ppm of HCN in leaf is dangerous to feed to cattle. If sorghum is manured heavily with N fertilizers, free NO_3 accumulates quite considerably and causes poisoning in the livestock. The NO_3 is reduced to NO_2 by microorganisms of the soil and rumen to convert haemoglobin to mathemoglobin which prevent transport and release of oxygen by the blood.

High quality forage is obtained from sorghum and legumes mixtures under good fertilizer management and package of practices.

Fodder maize

The chemical composition of the forage are as follows.

Crude Protein	-	6.74%
Ether extractives	-	2.09%
Crude fibre	-	35.95%
Total ash	-	8.15%
NFE	-	47.07%

Digestibility is 57 to 61% for dry matter, 22 to 67% for crude protein, 45 to 83% for NFE and 39 to 72% for crude fibre.

Cowpea

The feeding value of cowpea forage is high and comparable to Lucerne. It is superior to soybean because of its low fibre content and minimum wastage in feeding livestock. The chemical composition of the crop in % dry matter at harvest is as follows.

Crude Protein	-	15.77
Ether extractives	-	2.39
Crude fibre	-	20.11
Total ash	-	16.94
NFE	-	30.43

It is used as fodder crop for green feeding, making hay, grazing and also for ensoling in mixtures with sorghum or maize, cereal fodders in order to enrich it with crude protein and mineral contents.

Velvet bean

The forage contains 15% crude protein which varies with the stage of harvest and it has about 64% TDN. It is rich in Ca, P, Fe and I.

Cluster bean

The HCN content of the forage ranges from 40 to 70 mg/100g in early stages. This decreases with advancement of age and is almost in traces at maturity. It is not harmful at fruiting stage. The forage is palatable and rich in crude protein and minerals.

The composition is as follows.

Crude Protein	-	14 to 19%
Ether extractives	-	1.26 to 2.28%
Ash	-	8 to 17%

Calcium	-	1.56 to 3.65%
Phosphorus	-	0.1 to 0.4%
NDF	-	39 to 50%
Cellulose	-	25 to 33%
Lignin	-	5 to 11%

The digestibility coefficient of crude protein is high and varies from 70 to 85%.

Berseem

It is highly nutritious, succulent and palatable fodder for all types of livestock. It is rich in Ca and P.

Composition of green fodder is as follows :

Moisture	-	84.9%
Crude Protein	-	2.8%
EE	-	0.4%
CF	-	2.9%
NFE	-	6.6%
CaO	-	0.4%
P_2O_5	-	0.07%
K_2O	-	0.772%

Digestibility of crude protein is 81%.

Hybrid Napier or Hybrid-Pennisetum

The chemical composition is as follows (in percentage dry wt).

CP	-	10.2
EE	-	2.1
CF	-	30.5
Ash	-	16.2
NFE	-	41.0
Ca	-	0.5
P	-	0.4

The oxalate content of some of the varieties may be high (>3%). It can be mitigated if harvested at longer intervals (45 to 60 days).

Guinea grass

It is palatable to all kinds of livestock. The crude protein ranges from 4 to 14%. The crude protein content declines with the age of the herbage. The crude fibre content ranges from 28 to 36%. The EE content from 0.6 to 2.8% and NFE from 40 to 50%. The digestibility of crude fibre varies from 40 to 70% of TDN from 38 to 61% and of crude protein from 15 to 73%.

Para Grass (Buffalo grass)

The grass is poor in nutritive qualities.

The composition is as follows :

CP	-	3 to 16%
EE	-	0.9 to 3.9%
CF	-	28 to 34%
NFE	-	40 to 57%
Calcium	-	0.35 to 0.8%
Phosphorus	-	0.32 to 0.76%

Digestibility ranged from 39 to 6.3% for dry matter, 17 to 81% for crude protein , 40 to 78% for crude fibre. TDN content ranged from 41 to 71%. The nutritive value of the forage is improved if the grass is sown in mixture with legume.

Doob grass (*Cynodon dactylon*)

It is one of the most nutritious grasses. It is palatable and acceptable to all types of livestock. The dry matter constitutes 25 to 30% of fresh herbage. Crude protein content slowly decrease with plant age and the leaf: stem ratio is high. The digestibility for dry matter ranges from 45 to 60%, crude protein between 50 and 77%, crude fibre between 25 and 35%.

Anjan grass (*Cenchrus ciliaris*)

The crude protein content ranges from 3 to 16%. The calcium (CaO 0.5 to 2%) and Phosphorus (0.4 to 1%). Content lie in a well balanced ratio. EE 1.72%, Crude fibre 30.54%, Ash 16.08% and NFE is 43.3%. It is highly palatable and relished by animals.

Lucerne

The nutritive value in terms of percentage dry matter is as follows.

CP	-	19.9
EE	-	1.81
CF	-	29.51
NFE	-	34.68
CaO	-	2.8
P_2O_5	-	0.74
Ash	-	14.1
MgO	-	0.44

It is rich in Vitamins A and D, leaves constitute 45% of the dry matter and 80% of the protein remain in leaves. In Lucerne hay, the fibre content is low but the protein and digestibility are very high. It is particularly rich in calcium but poor in phosphorus.

Leucaena (Soo-babul)

The foliage contains an uncommon amino acid-mimosine which is toxic to non-ruminants at levels of about 10% of the diet. It is not fatal to ruminants since stomach micro-organisms convert mimosine to 3,4 dihydroxypyridine (DHP), which only on excessive and prolonged feeding results in ill health, poor growth, excessive salivation etc. swollen thyroids (goiters) are common among cattle, for prolonged feeding on leucaena. This is due to under production of thyroxine in the thyroid gland. Wilting the forage for 4 to 6 hours in the sun reduces its toxic effect considerably. Soaking of soo-babul in the water over night reduces the toxicity.

The leaflets contain 27 to 34% protein, which is of high nutritional quality. The fodder is a rich source of carotene and vitamin A.

The composition of the foliage is as follows.

CP	-	26%
Ash	-	11%
CF	-	20.4%
Calcium	-	2.36%
Phosphorus	-	0.23%

Contd...

Composition (%) of forage crops

Forage crops	Dry matter	Crude protein	Neutral Detergent Fibre	Cell contents	Acid Detergent Fibre	Hemicellulose	Acid Detergent cellulose	Cellulose	Silica
Berseem	17.00	16.90	53.00	47.00	39.95	13.05	12.55	26.85	0.55
Cow pea	14.50	17.50	50.00	50.00	37.90	12.10	9.00	27.00	1.90
Lucerne	22.40	17.80	46.00	54.00	32.00	14.00	9.60	21.50	0.90
Bajra	19.50	8.20	67.90	32.10	38.30	29.60	7.80	27.70	2.80
Jowar	31.62	5.90	70.70	29.30	44.42	26.28	7.55	34.15	2.72
Maize	28.80	6.70	66.40	33.60	38.45	27.95	6.80	28.55	3.10
Paragrass	36.50	8.50	63.95	36.05	29.20	34.75	4.22	21.85	3.12
Sudangrass	24.81	6.55	69.80	30.20	44.25	25.45	9.45	32.40	2.50

Chemical composition and *in-vitro* nutrient digestibility (% of dry matter) of forage crops at different stages

Forage crops	Dry matter	Crude protein	Neutral Detergent Fibre	Cell contents	Acid Detergent Fibre	Hemi cellulose	Lignin	Cellulose	Silica	Ash	In Vitro DMD	In Vitro CWD
Legumes												
Pre flowering	14.67	23.36	41.07	58.93	34.01	7.06	9.61	23.61	0.77	12.22	73.07	34.81
Flowering	18.23	18.90	48.84	51.16	39.39	9.45	11.15	27.40	1.26	10.71	66.40	31.32
Post harvest	28.08	15.68	54.12	45.88	43.17	10.95	12.27	29.28	1.52	9.60	63.58	33.33
Non-Legumes												
Pre flowering	15.50	10.49	55.16	48.84	36.10	19.06	5.81	28.36	1.74	10.61	70.10	45.93
Flowering	24.20	7.86	62.45	37.55	40.21	22.24	6.96	30.74	2.29	9.69	63.60	42.69
Post harvest	34.37	6.13	69.45	30.55	44.62	24.83	8.01	32.23	2.91	9.46	57.40	39.06

Chapter 16

Post Harvest Changes in Pulses

Insect and rodent infestation and mould growth during post harvest handling, storage and distribution of pulses, causes substantial losses, both quantitative and qualitative. Birds and rodents accounted for considerable losses during drying and storage of pulses. Products of pea were found to be heavily infested. Pulse beetle was reported to cause heavy destruction of pulse seeds in addition to making them unhygienic, because of the presence of excreta. Seeds infested with these pests, were unfit for human consumption, as these were contaminated with metabolic such as uric acid and had increased fat acidity. Bengal gram seeds infested with pulse beetle during storage showed changes in their nutritional characteristics and were found to be inferior. Insect infestation significantly decreased the nutritive value of bengalgram and pigeonpea, proteins, and reduced their milling yield.

A single treatment by a combination of fumigation and pest-proofing of bags was reported to preserve the seeds in sound condition for about six months. But it should be possible to develop grain legume varieties with increased resistance to post-harvest infestation to alleviate storage problems and to reduce the need of chemical treatments for the control of insect infestation and microbial infection.

❑❑❑

Chapter 17

Post Harvest Changes in Oilseed and Sugarcane

Environmental effects on the fatty acid composition on oil seed lipids

The fatty acid compositions of a number of seed oils have been shown to vary with the temperature at which the seeds have developed with lower temperatures favouring more unsaturated oil. Temperature changes which in the long term results in alterations to the fatty acid composition of linseed and soybean oil also induce similar modifications to the fatty acid compositions of PC (Phosphatidylcholine) and diacylglycerol in the developing cotyledons of these species, but much more rapidly. The extent of stearate conversion to unsaturated acids changes with temperature, suggesting that effect is mediated by the solubility of oxygen in water, which is temperature dependent. However, the major change in seed oils occur between oleate and poly unsaturated fatty acids (linoleate or linolenate) indicating that the olicate desaturate rather than the stearate desaturate is important in determining the temperature response in oil seeds.

Increase in free fatty acid content

The increase in fatty acids in seeds is largely due to invasion by fungi and is a major symptoms of deterioration only at seed moisture contents above twelve per cent. Fungal invasion is a major cause of break down of lipid to free fatty acids. Fungal invasion becomes serious in seed stored at or above moisture levels of fourteen to fifteen per cent, which is in equilibrium with relative humidites of about seventy five per cent and optimum temperature for fungal invasion is 30° C.

Lipid auto oxidation

It has been shown that free radicals of lipid peroxides damage cytochrome-C by changing its physical and catalytic properties, suggesting that free radicals attach themselves to the protein by covalent linkages. It most often affects poly unsaturated fatty acids, because they contain multiple double bonds in between which lie methylene bridges ($-CH_2-$) that posses reactive hydrogen.

Oxygen reacts with many organic substrates to yield hydroperoxides and other oxygenated compounds.

$$RH \xrightarrow{\text{Initiator}} R$$

This production of free radicals may take place by direct thermal dissociation (thermolysis), by hydroperoxide decomposition, by metal catalysis and by exposure to light (photolysis).

Propogation

$$R + O_2 \rightleftarrows ROO$$

$$ROO + RH \rightarrow ROOH + R$$

Auto oxidation depend on the availability of allylic hydrogen for reaction with peroxy radicals

$$ROO + R - CH_2 - CH = CH - R' \rightarrow R - CH - CH = CH - R' + ROOH$$

Termination

$$ROO + ROO \rightarrow \text{non radical products}$$

The most important termination process for secondary peroxy radicals at room temperature involve a tetroxide intermediate to produce ketone, an alcohol and oxygen.

Lipid auto oxidation is accelerated at high temperature and increased oxygen concentration. This was considered as a cause of seed deterioration only at moisture contents below six per cent.

Post harvest changes in sugarcane

Post harvest deterioration of cane occurs mainly due to delay in crushing of the harvested canes. The delay could be either in transporting or may even in the yard.

When cane is harvested and stored, loss in cane weight due to moisture loss and bio deterioration due to leuconostoc bacteria occurs. The moment cane is harvested the leuconostoc bacteria enters the cut ends, and it has been found that within one hour after harvest, the bacteria can pass upto 20 cm in the stem from the cut ends. The bacteria causes degradation of sucrose with the formation of dextrin and other gummy materials. Maximum deterioration of the harvested cane occurs at 38 to 40°C at 15% moisture loss. The loss in cane weight depend upon the maturity status of the crop and the prevailing weather conditions. In the tropical region of India, the loss has been found to range from 1.5 to 2.3% for every 24 hours storage after harvest during the season. Higher loss upto 25% have been reported during hot season. In the subtropical belt, the loss is lower during winter and higher during hot period i.e., late harvested canes. Apart from losses in cane weight and sucrose per cent juice, deteriorated cane adds to reduced juice extraction, there will be problem of clarification and filtration. A dextrin level of above 500 ppm results in the formation of needle shaped sugar crystals with difficulties in the separation of molasses. The rate of deterioration and loss in sucrose per cent is very high when the time lag exceeds 48 hours after harvest. The deterioration of the juice quality is much faster when the canes are cut into number of bits.

❑❑❑

Chapter 18

Post Harvest Changes in Fruits and Vegetables

Physiological and biochemical changes in harvested vegetables

Vegetables contain more than 80% water and some vegetable such as tomato, cabbage and cucumber contain as high as 90 to 95% water. As little as 5% loss in water causes many vegetables to appear wilted or shriveled which make the vegetable tissue tough or musty, losing, crispness and palatability and eventually unmarketable. This happens to several leafy vegetables and root vegetables within a few hours after harvest. Water loss is also associated with degradative effects on other vegetable components such as pigments and vitamins.

Vegetables exhibit water loss mainly by transpiration through stomata. Structure of vegetable and the physiological age at harvest largely influence the rate of water loss which in turn depends upon the relative humidity, temperature, air movement and atmospheric pressure of the environment. In general, larger the vegetable surface to its volume, more rapidly vegetables loss moisture.

The respiratory activity increased 1.7 times at 5°C, 2.8 times at 10 °C, 4.5 times at 15 °C and 6 times at 20 °C, physiological maturity at harvest, structure of the peel, volume of inter-cellular spaces, and also the mechanical injury during handling have influence on the respiratory activity of the produce to be stored or marketed. Respiration primarily involves enzymatic oxidation of sugars to CO_2 and water with release of energy. Loss of other compounds such as proteins, lipid and organic acids is also channelized through respiration. Higher the rate of respiration shorter is the shelf-life. It results in the reduction of reducing sugars, non reducing sugars and total sugars. Changes in respiration, transpiration, weight, pigments, organic acids and flavour vary in individual vegetable. Colour changes occur due to breakdown of the pigments present at harvest and also due to synthesis of new pigments after harvest. Relative changes in the proportion of sugars and organic acids lead to characteristic flavour. The vegetable which lost water more rapidly and wilted appreciable tended to lose vitamin C more rapidly than those resistant to wilting.

Ethylene causes serious disorders in leafy vegetables at very low concentration. The diseased or mechanically injured specimens of vegetables produce more ethylene, causing excessive senescence of the vegetables tissue and storage of such specimens with other vegetable should be avoided. In fruit vegetable such as tomato, ethylene causes fast ripening. Sprouting of tuberous and bulbous growth, amounts to loss of the commodity as a vegetable. Development of toxic pigments such as solanine in potato, on exposure to light, is also a characteristic unwanted post harvest process in vegetables.

Nature and causes of losses

Owing to the high moisture content and tender nature, vegetables pose characteristic post-harvest problems.

Primary causes of losses

a. Mechanical losses

Besides causing damage bruising and cracking make the vegetable more prone to attack by organisms and significantly increase the rate of water loss and gaseous exchange.

b. Physiological and biochemical losses

In fresh vegetables transpiration and respiration and sprouting of tuber and bulb vegetables lead to direct food loss. The loss is accounted towards senescence process and represents major components of post-harvest losses. Post-harvest rooting in tuber vegetables, seed germination, greening of potatoes leading to production of harmful compounds and toughening and spongyness in carrot and radish that represent physiological loss, lower down the quality of vegetables.

Microbial losses

A significant portion of losses of vegetables during post harvest is attributed to diseases caused by fungi and bacteria. Succulent nature of the vegetables make them easily invadable by the organisms. Besides attacking fresh vegetables, the organisms also cause damage to canned and processed products.

Physical losses

Excessive or insufficient heat during processing, improper cold storage temperature and relative humidity and undesirable gaseous composition of controlled atmosphere storage lead to physical damage due to tissue break down.

Secondary causes of losses

Inadequate harvesting, transportation, storage and marketing facilities and legislation lead to conditions favourable for secondary causes of losses.

Control of post-harvest losses

Cultural operations

Cultural operations which ensure normal development of vegetables and avoid infection of decay organisms, help to prolong shelf-life. For root crops such as carrot and radish, preparation of the soil to a fine filth of porous is necessary to avoid root forking. Regular irrigation during development of bulb, tuber and fruit vegetables is of vital importance.

Harvesting

Consumers always prefers fresh, tender and disease and insect free vegetables of attractive appearance. Vegetables should be harvested as and when they attain maximum size and yet are tender. Vegetables other than root crops should not directly be placed on the soil.

Pre-harvest treatments

Post-harvest applications of metabolic inhibitors such as maleic hydrazide reduces sprouting of onions and potatoes during storage and those of growth promoters such as N benezyladenine prolongs the shelf-life of leafy vegetables. Ethylene induces early maturity.

Washing and grading

Root and tuber crops are often washed to remove adhering soil. Besides cleanliness, water used for washing improves vegetable appearance and prevent wilting. Vegetables are graded according to shape and colour, size, maturity and general appearance as well. In sorting the vegetables into different grades, all characters that influence appearance and quality of the produce should be considered.

Packaging

Cheap packaging techniques and material such as polyethylene films, paper board boxes lines with polyethylene and other material can effectively prolong the shelf-life of vegetables.

Curing

Curing of onions in crates of 40°C for 16 hours is found to reduce rot losses in storage. In other crops such as potato, garlic, sweet potato cassava, yam and others curing effectively reduces post-harvest decay and water losses.

Low temperature storage

Refrigeration or low temperature storage can most effectively extent shelf-life of vegetables and reduce post-harvest losses by arresting metabolic break down and fungal deterioration of the commodity.

Effect of Temperature and Relative Humidity on the storage life of vegetables

	Vegetable	Temperature	RH (%)	Storage life (Weeks)
(i)	Brinjal	10 to 11.1	92	2-3
(ii)	Bitter gourd	0.6 to 1.7	85-90	4
(iii)	Cabbage	0.0 to 1.7	92 to 95	4-6
(iv)	Ginger	7 to 10	75	16-24
(v)	Onion bulbs	0.90	70 to 75	20-24
(vi)	Tomato ripe	7.2	90	1

Controlled atmosphere storage

The controlled atmosphere storage wherein higher CO_2 and lesser O_2 are maintained. It retards respiratory activity, delay softening, yellowing, quality changes and other deteriorative processes.

Ionizing radiation

It inhibits sprouting of potatoes, onion, and other bulbs most effectively at 10-15 krads. One of the significant advantage of the ionizing radiation is that it can destroy pathogenic organisms without appreciably raising the temperature of the produce.

Post harvest treatments of chemical

Wide range of chemicals have been used to control microorganisms (fungicides, antibiotics), ethylene level (ethylene absorbents), water loss (antitranspirants) and senescence (growth substances) desiccants' (calcium chloride, fumed silicas) for dehydrated products also give added advantage.

Post harvest changes of fresh fruits

Physiological and biochemical aspects

Fruits after harvest lead an independent life as they are removed from the parent plants and the normal supply of water, minerals and other organic molecules are completely cut off. The harvested fruits do not have any significant photosynthetic process but they carry out all other physiological processes such as transpiration, respiration, ripening, etc.

Maturity

The post harvest quality and storage life of fruits appear to be controlled by the maturity, if the fruits are harvested at a proper stage of maturity the quality of fruit is excellent. Poor quality and uneven ripening are caused by early harvesting. Maturity can be described as the attainment of a particular size or stage after which ripening takes place. On the other hand ripening means the qualitative changes in fruits after maturity, as a result of which it becomes edible.

Various maturity indices such as number of days from fruit-set, visual indicators, size, shape, colour, appearance, texture, specific gravity, starch content, soluble solids, sugar : acid ratio and oil content are used to determine the maturity of fruits.

In mango variety Alphonso it took 110 to 125 days after fruit-set for surface colour to change from dark green to olive green and fresh colour from white to pale yellow. It was also observed that fruits having specific gravity above 1.04 were partially ripe and below that were green at the time of harvest. The maturity of pineapple is characterized by accelerated decline in shell pH and marked increase in fresh brix and titratable acidity. Starch is absent in mature orange fruits.

Respiration

During respiration the sugar disappears with the uptake of O_2 and production of CO_2 and H_2O. A great deal of energy is liberated as heat at the time of respiration. The initial phase of respiration is production of pyruvic acid from glucose, a phase known as glycolysis. The terminal phase of respiration is the conversion of pyruvic acid into CO_2 and H_2O which is strictly aerobic. In most of the fruits, the rate of respiration increases rapidly with ripening. The sudden upsurge in respiration is called the turning point in the life of fruits. After this, senescence and deterioration of the fruit being take place. The fruits belonging to this type of respiratory process are called climateric fruits. In nonclimateric fruits, there appears to be a simple gradual decline in respiration. Banana, papaya, mango, sapota etc. are the major tropical fruits which belong to climateric group and pineapple, grape, lemon, etc. belong to non-climateric group. The rate of respiration is a good index of the post-harvest life of fruit. To extend the post-harvest life of the fruit, its respiration rate should be reduced as far as possible. Once harvested, carbohydrates utilized during respiration cannot be replaced and the calorific value of the fruit is reduced.

Ripening

With ripening, changes occur in colour, flavour and texture which render the fruit acceptable to the consumer. These quality changes are accompanied by respiratory changes in the fruit and compositional changes such as transformation in pigments, pectins, carbohydrates, acids, phenolics, etc. colour change in fruit is due to degradation or breakdown of chlorophyll and consequence unmarking of carotenoid pigments as in case of oranges. Fruit softening is caused by the break down of insoluble protopectin and formation of soluble pectin. During ripening there is decrease in starch and increase in fructose, glucose and sucrose resulting in sweetness of the fruit. Ripening also decreases the organic acids and phenolics to reduce the acidity and astringency and makes it more acceptable. An increase in volatile compounds to give the characteristic flavour of a fruit also takes place during ripening.

The most striking chemical changes which occur during the post-harvest ripening of banana are the hydrolysis of starch and accumulation of sugar.

Ethylene

It is known as the 'ripening hormone'. The amino acid methionine is the immediate precursor of ethylene production in fruit tissue. Mature unripe fruits are stimulated to ripen if treated with very low concentration of ethylene gas of about 0.1 ppm. Application of ethyl

(2-chloroethane phosphenic acid) imparted attractive colour to mangoes and increased their commercial value but reduced the ascorbic acid content.

Advancement of ripening by application of Ethrel has also been observed in many fruits by various workers. Fruits ripened by calcium carbide had better palatability, attractive colour and superior quality attributes. Citrus fruits are degreened with Ethephan which actually wider carotenoid synthesis, apart from hastening chlorophyll degradation. Application of ethylene at 0.1 ppm or higher to green bananas accelerate the onset of the climateric stage. Ethylene is 100 times more effective than acetylene as a ripening agent. During the ripening of alphonso mango ethylene gas evolved was in the range of 0.02 to 0.18 ppm. There are various ethylene liberating compounds available for senescence.

❑❑❑

Chapter 19

Classification of Pesticides

Classification of Pesticides

Pesticides, the plant protection chemicals, being potential toxic substances by very nature, warrants sufficient care to safeguard the users, animals, birds and wildlife and the environment at large from the associate hazards and environmental pollution. A chemical to be used for pest control cannot be merely any toxic substance that can kill the pest and/or disease. Many factors like phytotoxicity, effect of the pesticides on animals, pollution effect, degree of persistence of the residues of the chemical on the crops have also to be considered while selecting any chemical for use. Chemicals used for controlling pests can be divided into different groups based on the nature of pests to be controlled like

Acaricides	-	for the control of mites, ticks, etc.
Algicides	-	for the destruction of algae and aquatic vegetation
Antiseptic	-	for protection of non-metals from micro-organisms
Arboricides	-	for the destruction of undesirable arborell and bushy vegetation
Bactericides	-	control of Plant pathogenic bacteria and bacterial diseases
Fungicides	-	for the control of various fungi
Herbicides	-	for the control of weeds
Insecticides	-	for the control of harmful insects
Limacides or Molluscides	-	for the control of mollusics including gastropods, snails and slugs
Nematicides	-	for the control of nematodes
Rodenticide	-	for the control of rodents

Chemicals used for stimulating or retarding the growth of plant (growth regulators), for the removal of leaves (defoliants), for desiccating plants (desiccants), for repelling insects (repellents), for attracting insects (attractants) and for sterilizing insects (sex sterilants) are also included among the pesticides. However, out of these, insecticides, fungicides and herbicides in that order are most widely used.

Classification of Insecticides

A. Based on mode of entry

Based on the manner in which toxicants are administered to the insects and their mode of entry into the body, the insecticides are classified into.

(i) Stomach poison

The toxicant applied to the food, when ingested by the insect kills it primarily by action on or absorption from the digestive system. Though usually it is limited to the control of chewing insects, under, certain conditions it may also find use in controlling insects with sucking, siphoning, sponging or lapping type of mouth parts.

(ii) Contact poison

A toxicant which kill the insect by contact and entry into the body through the vulnerable sites found on its body is said to be a contact poison. It may be directly applied on to the body of the insect in a spray or dust.

(iii) Fumigant

A toxicant applied as a vapour enters the tracheae of the insect through the spiracles in the form of gas and brings about its kill, such a gaseous toxicant is called a fumigant.

B. Classification based on mode of action

Based on the ways in which the chemicals act upon the system of an insect to cause its death, insecticides may be classified as follows:

(i) Physical poison

A physical poison exerts a physical rather than a biochemical effect and brings about death of insects by asphyxial effect (exclusion of air) as with heavy oils and tar oils or by effecting a loss of body moisture from the insect by inert dust.

(ii) Protoplasmic poison

A protoplasmic poison is primarily associated with precipitation of protein. The cellular protoplasm of midgut epithelium is destroyed by inorganic stomach poisons.

(iii) Respiratory poison

It is associated with blocking of cellular respiration and in activation of respiratory enzymes as with fumigants.

(iv) Nerve poison

The action is primarily associated with the solubility of the toxicant in tissue lipoids and inhibition of acetylcholinesterase in insects and warm-blooded animals.

(v) Poisons of a more general nature or action

Aldrin, chlordane, dieldrin and toxaphene do not induce neurotoxic symptoms until a latent period has passed.

C. Classification based on chemical nature of insecticides

Two major divisions, viz., inorganic and organic insecticides are recognized.

Inorganic compounds

- Arsenicals
- Fluorine compounds
- Sulphur and lime sulphur
- Zinc phosphide

Organic compounds

a. Hydrocarbon oils
   - Oils (Petroleum or mineral oils)
   - Tar oils

b. Animal origin-Nereistoxin

c. Plant origin
   - Nicotine
   - Pyrethroids
   - Rotenoids
   - Volatile oils from plants

Synthetic organic compounds

- Dinitrophenols
- Thiocyanates
- Organochlorines
- Organophosphorus compounds
- Carbamates
- Sulphur containing compounds
- Mixed oils and soaps
- Fumigants
- Nuclear polyhedrosis (NPV)

❑❑❑

Chapter 20

Formulations of Pesticides - Dusts and Wettable Powder

Formulations are classified into the following groups based on the form in which each material is utilized or the purpose for which it is employed.

Dry formulations

- Dusts
- Granules
- Wettable powders
- Seed disinfectants
- Others

Dusts

They consist of a mechanical mixture of the active ingredients with or without inert diluent, pulverized to a particle size of 3 to 30µ. They can be classified as follows.

Undiluted toxic agent	- Sodium fluoride
Toxic agent with an active diluent	- Rotenone with sulphur
Toxic agent with an inert diluent	- Lindane with pyrophyllite
Inert dusts	- Silica

Insecticides which are to be used dry are mixed with organic flours such as walnut-shell flour, soybean and wood bark or pulverized minerals such as sulphur, lime, gypsum, talc, etc. The concentration of toxicant present in the dust formulation ranges from 0.1 to 25 per cent. They are ground to fine sizes ranging from 1 to 40 microns and most dust particles will pass through a 325 mesh screen. In general, the toxicity of a dust formulation increases as the particle size decreases. The quality of dust formulation largely depends on the properties of the carriers used. Compatibility, particle size, abrasiveness, absorbability, specific gravity, wettability and cost are some important basic points to be considered in the selection of dusts. Dusts are easy to handle. Best results are obtained if applied early in the morning when the plants are wet with dew. Even slight wind may cause drift of dust to other fields.

Granular or pelleted insecticides

Insecticides are formulated as granules consisting of inert materials with the toxicants 0.25 to 2.38 mm in diameter and the range of particle size is designated by a two-figured mesh classification such as 30/60 which means that virtually all of the granules will pass through a standard 30 mesh sieve while only a negligible quantity will pass through a standard 60 mesh sieve. The formulations commonly contain 2 to 10 per cent toxicant applied by solvent impregnation to highly absorptive clays such as benetonites and attapulgite. Granules are broadcasted either on to the plant or soil and no water is required for their application. There is very little drift which prevents undesirable contamination and undue loss of insecticide. Granules do not adhere to plant surface and thus decrease residual deposits. The toxic material in the granular formulation is released over a longer period than does a spray deposit. Granular application may cause scorching sometimes if the chemical is concentrated in a smaller volume of carrier. The efficiency of the granular formulation applied to soil is dependent on many factors such as dosage, type of formulation, soil type, moisture condition, method of application, rate of release of the toxicant, type of crop plant, weather conditions, etc.

Wettable powders

Powdered formulations that give stable uniform suspensions with water are called wettable powders. Suspensions are more effective than dusts. Suspensions are preferred because the particles of a suspension adhere well to the surfaces and also do not penetrate and can be washed out of required. Important conditions for the wettable powders to be effective are:

- That they should be stable in storage and should not cake.
- The suspension should form rapidly but the solid particles should settle easily.
- The sprayed objects should have good wettability and also the sprayed material.

- Should spread easily over the surfaces, and
- The sprayed material should be retained on the sprayed surfaces for a long time.

Wettable powders should have a high degree of dispersion of the active ingredient of the preparation. Good wettable powders usually contain not less than 80 per cent of particles with a size below 3 μ. Wettable powders can be divided into three groups

(i) Preparation with a high content of the active ingredient content (>6%)

(ii) Preparations with a medium content (30 to 60%) and

(iii) Preparations with a low content (<30%).

In a high active ingredient content formulation, pure materials without oily impurities and a diluent with a low bulk density and relatively large sorptive capacity are used. Wettable powders without diluent can be prepared if the active ingredient has a high melting point and grinds well. With a poor quality diluent, the quantity of surface active agents have to be increased as otherwise unstable suspensions will be obtained. High concentrations wettable powders are produced by grinding in a colloidal mill a mixture of the active ingredients with the diluent and the other ingredients. In the case of pesticides obtained by precipitation from aqueous solutions, all materials required for the preparation of the wettable powders are added to the precipitated and washed pesticide and then the mixture is dried in spray driers, thus avoiding the grinding. Wettable powders such as copper oxychloride, ziram and cuprous oxide are prepared. Wettable powders with a compound action have also been prepared.

❏❏❏

Chapter 21

Formulations of Pesticides - Emulsion Concentrates and Water Miscible Liquids sprays

Sprays

In places where water is available, liquid formulations are economical, efficient and convenient to apply. Unless the insecticides are soluble in water, they are usually formulated as emulsifiable concentrates in an aromatic solvent such as zylene or in a petroleum oil. The choice of a solvent is governed by the degree of solvency, compatibility, cost and availability. With easily hydrolysed insecticides, the water content of the solvent must obviously below. These concentrations are prepared by dissolving the chemical in the solvent with emulsifiers of a type and proportion to provide the desired properties of suspensibility and emulsion stability under conditions of field use. Emulsifiers commonly used are, nonionic with anionic surface agents. The nonionic types are generally insensitive to hardness of water and store well. The anionic types produce low interfacial tension thus favouring the formation of small dispersed droplets in an easily emulsified mixture. Most of the surface active agents or surfactants may react with acids to form complexes which are no longer surface active and are therefore devoid of emulsion promoting properties. Soaps may also be used as emulsifiers but only when the water to be added is soft.

Emulsions may be undesirable when two or more formulated sprays (a fungicide and an insecticide) are to be mixed. Here while the actual pesticide may be compatible, either the solvent or emulsifiers may not be compatible. Good coverage of the sprayed surface should be there. In addition to the emulsion stability in spray tank, it should break down quickly as soon as it hits a foliage or other oily or waxy surfaces. The main disadvantages of dilute spray is the enormous quantity of water needed for spraying, in many cases upto 900 litres. With low volume nozzles, and ultra low volume nozzles, the same quantity of pesticides could be sprayed in 65 to 90 litres of water or even in 10 litres of water over a hectare. However the formulation for LV or ULV have to be slightly different from the regular emulsion concentration. Liquid insecticides may be applied as water or oil solutions, or water suspensions of solid particles or water based emulsions.

Solutions and solubility

Petroleum distillates and water are the most common solvents of insecticides. Materials which are insoluble in both require some intermediate or mutual solvent often called a solubilizer, blending agent or coupling agent. Pyrethrin although insoluble in water, may be dissolved in petroleum and then emulsified in water. If unsaturated compounds are used as solvents it is possible that they may react with even such stable compounds as DDT and the duration of the efficiency of the preparation will be less, especially in light and at high temperatures sometimes it will be necessary to mix two or more solvents to increase the solubility, for example the pesticide may be first mixed with small amount of expensive solvent and then the cheaper solvent is added. The first one used is called intermediate solvent. Intermediate solvents for the preparation of concentrated solutions of pesticides in mineral soils are: cyclohexane, methyl oxide, thiophene, methyl ethyl ketone and chlorobenzene. Intermediate solvents for halogenated and organophosphorus insecticides are carbon tetrachloride, dichloroethane, trichloromethylene and tetrachloroethylene. Most frequently used solvents for preparation of pesticide solutions are petroleum hydrocarbons, mineral oils and diesel.

Wetting

Wetting of a solid occurs when a liquid comes in direct and persistent contact with the surfaces of the solid and no layer of air separates solid from the liquid.

Spreading

Spreading of a liquid on a solid surface is defined as the ability of the liquid to form a persistent liquid-solid interface, slowly by the surface activity over the solid surface.

Emulsifications

To reduce oil deposits and thus avoid diluted with water. As water is not miscible with oil directly, various processes of emulsification have been developed to disperse the oil uniformly in water. As each oil droplet is surrounded by a more or less tough film of emulsifier, the tendency of droplets to coalesce and separate into a continuous layer of oil is resisted. Two types of emulsions exist. i) oil in water or OW type; water is external or continuous phase and oil is the internal or disperse phase. Water in oil or WO type; oil is external and water is internal. Because an emulsion can be diluted only by adding more of the external phase, the OW type of emulsion is the one almost universally employed for insecticidal purposes.

Emulsifiers

Soaps, sourmilk, glue and dry skim milk which are colloidal substances have been used as emulsifiers. Alkali soaps are the most commonly used emulsifiers.

Deposit builders

The addition of a spreader often increases the drain of spray materials from the foliage, but the remaining insecticide forms a more uniform film type of deposit. The sticker may act indirectly by causing flocculation of the spray particles. Additives which increase the initial deposit appear to be of a type such as gelatin, glue and casein in preparations which increase the viscosity of the spray.

Adhesives

Where persistence of deposit or residual effectiveness is required, adhesives should replace spreaders. Proteins and similar colloids tend to adhere and if adsorbed on the particles of insecticides tend to adhere and if adsorbed on the particles of insecticides, will promote adhesion. Inorganic materials having hydrophilic properties are adhesives. Materials of adhesive value included a wide variety of substances like molasses, sugar, dextrin, starch, wheat flour, soybean flour, gum, gelatin, casein, milk, glue, tannins, egg powder, etc.

Emulsive concentrations

The emulsive concentrates upon dilution with water give stable emulsions for spraying plants and surfaces. Two kinds of emulsive concentrates are known.

(i) Emulsive concentrates prepared by dispersing an aqueous solution of pesticides in a water immiscible solvent by means of a colloidal mill. These concentrates are more properly called concentrated emulsions". Highly dispersed concentrated emulsions can be obtained that are stable in storage, by mechanical dispersion. Homogenised concentrated emulsions withstand low temperatures and prolonged storage without substantial changes.

(ii) Emulsive concentrates of pesticides known under the name of miscible oils; when the miscible oils are mixed with water, they yield stable emulsions. Concentrates of this type consist of a pesticide, solvent and emulsifier.

Hydrocarbons, halogenated derivative of hydrocarbons, esters, various petroleum products, kaolin, coal tar oils may be used as solvents. Emulsifiers are calcium sulfonates, esters of polyethylene and polypropylene glycols, various soaps and salts of naphthalenic acids. Good results are obtained by the use of mixtures of two or more emulsifiers. It is convenient to have concentrates that are not too viscous. However for highly toxic organophosphorus compounds it is less dangerous to use viscous concentrates because if they get on the skin, they are more difficult to absorb and can easily be washed off with water. There is a wise use of the so called "inverted emulsions" that are also prepared by dilution of the appropriate concentrates with water-in-oil type is obtained. These concentrates are used for low volume spraying since the presence of an oily film hinders evaporation of the drops. To decrease the evaporation of drops in aerial spraying, special additives like LOVO (butylamine state) are introduced. As the pressure of spraying increases, the droplet size, impingement per unit of surface area, time for spraying and mortality of insects are all reduced.

❑❑❑

Chapter 22

Formulations of Pesticides - Aerosols and Fumigants Insecticidal Aerosols

Insecticidal Aerosols

An aerosol is a system of particles of ultra microscopic size dispersed in a gas. Insecticidal aerosols are produced by either incomplete combustion of tobacco or other organic material containing an insecticide or by spraying a solution of an insecticide such as DDT in fuel oil on a heated surface or by spraying a solution of an insecticide from a container of liquefied gas.

Aerosols fogs and smoke are normally produced by a toxic aerosol released from a pressurized container by means of a propellent gas or by a heat induced volatilization of a particular insecticide in a suitable petroleum or other solvent. A fog is composed of liquid particles and a smoke is composed of solid particles. The simplest method of producing pesticide aerosols is burning in special pots, paper and combustible porous materials that have been impregnated with insecticides, fungicides and bacteriocides. Besides the pesticides, the combustible smoke forming compositions usually contain a combustible substances, filler and an oxidizing agent.

NO_3, NO_2, chlorates, chromates, bichromates, per sulphates, etc } Oxidising agents

Kaolin, infusorial earth filler.

With the aid of CO_2 or a low boiling solvent (Freon, methyl chloride) pressure is created in the cylinder that facilitates good atomization of the preparation. When the solution is sprayed into the air, the solvent will evaporate and the pesticide will remain in the atmosphere in a finely dispersed condition. The use of low volume finely dispersed spray is effective both in sheltered ground and for spraying large bodies of forest and fields with a small expenditure on the preparation.

Fumigants

Insect fumigants must be sufficiently volatile to produce a toxic concentration of vapour in closed space. Fumigant concentration is usually expressed in weight per volume. The vapourisation of fumigant chemicals seldom approach the point of saturation. The distribution of fumigant vapour should be uniform throughout the treated surface. Diffusion is faster with gases having lower molecular weights.

Molecular diffusion must be relied upon to bring about gas penetration of furniture, food and other treated goods. Among the hazards in handling insect fumigants, two should be stressed because of their insectisidious nature which may result in unwitting accidents like absorption by skin and inflammability of the fumigant-air mixtures. Substances absorbed through the skin included carbondisulphide, carbon tetrachloride, ethylene dichloride, Cl_2, I_2, phenol, etc.

Building fumigation is concerned mainly with the killing of stored food and fabric pests in infested hours, storage warehouses, factories and transportation facilities. Product fumigation involves the fumigation of such commodities as grains, flour and tobacco. Crop fumigation concerns mainly the gassing of citrus trees enclosed intents with HCN. Soil fumigation is used for killing of soil infesting insects, nematodes and root fungi.

Flowability

The flowability of a dust is an indicator of the feed rate of dusting equipment. The angle of slope which is a measure of flowability, is measured by allowing the dust to flow through a funnel upon a disc. Greater the angle the poorer is the flowability. For example, sulphur dust treated with a conditioner has an angle of 40° whereas untreated sulphur has angles $> 70^\circ$. Dusts with fibrous or needle shaped particles have a slower feed rate than dusts with spherical particles.

❑❑❑

Chapter 23

Organochlorine Insecticides

Organochlorinated Hydrocarbons

Mode of action

Two general theories based on Overton-Meyer theory on lipid solubility and physiological activity have been advanced to explain the relation of chemical structure of organochlorine compounds to insecticidal activity. Langer and his associates in 1944 postulated that the trichloromethyl part of a molecule is a requirement to bring about penetration to the site of action and that the P-chlorophenyl groups linked by a carbon alone constitute the toxic components of the molecule.

With regard to the chemical mechanism responsible for toxicity, Campbell and West suggested the evolution of hydrogen chloride as only a secondary effect accompanying the more fundamental alteration in the electromagnetic nature of the group from which it is evolved. The effect of chlorine atoms are therefore may be entirely steric and that the three atoms of chlorine are necessary for conferring optimum toxicity on the compound due to their providing the required steric orientation and that this reaches a maximum when the remaining chlorine atoms are in the para-positions. The active substituents in general must be in the para positions (4,4 positions) for the compound to have a high degree of toxicity. The presence of additional substituents in other positions appear to cause a diminution of activity. The insecticidal activity of organochlorinated hydrocarbons is related to the case with which it is absorbed by the insect cuticle.

They affect the peripheral sensory organs to produce violent strain of different impulses that produce hypersensitivity and then convulsions of the insect.

Lindane

BHC is prepared by chlorination of benzene.

$C_6 H_6 + 3Cl_2 \rightarrow C_6 H_6 Cl_6$

To obtain γ-isomer the crude product is treated with calculated amount of methanol at a particular temperature, and the undissolved α and β isomers are filtered off and the filtrate is cooled. By this process the technical g-isomer is isolated after removal from the mother liquor and dried. It contains > 90% of the g-isomer. To obtain lindane (99% of pure isomer) this preparation is again recrystallised and g-isomer obtained.

The efficiency of extraction and yield of pure Lindane depends upon the temperature, properties of various compounds and size of the crystals in the technical materials. The pure compounds was named as Lindane after Vander Lindane, the discoverer of gamma isomer. It is appreciably volatile at room temperature and sublimes most rapidly. It is widely used because it has no odour and also leaves no residues on food and forage products. Because of its cost, it has been widely used in household sprays, livestock dusts and seed treatment. It is colourless crystalline solid, practically insoluble in water but soluble in organic solvents. It is odourless. It is very stable to the action of heat, light and oxidation but is readily decomposed by alkaline materials.

Mode of Action

Lindane rapidly penetrates the insect inside and can exert a significant tangent action in a dry atmosphere.

Endosulfan

Thiodan, 6, 7, 8, 9, 10, 10 a-hexachloro 1, 5, 5a, 6, 9, 9a-hexahydro-6, 9-methano 2, 3, 4-benzodioxathiepin-3-oxide.

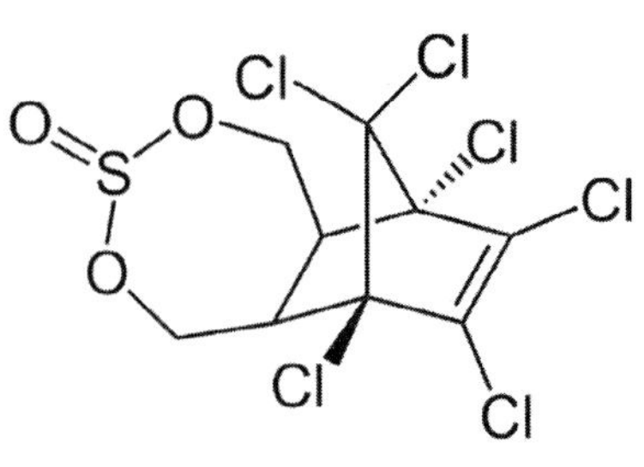

It is a broad spectrum insecticide for pests of fruits, vegtables and ornamentals.

It is the product of hexachlor cyclopentadiene and 1,4-dihydroxy-2-butene subsequently chlorinated with SO_2 / Cl_2 to produce the endosulfan. The technical material is a brownish solid, m.p. 70 to 100°C consisting of about 4 parts of a- cis isomer m.p. 180°C and one part of β trans isomer m.p. 206°C the a-cis isomer is slowly converted to the more stable β form at high temperature. Both the isomers are slowly oxidized in air and in biological systems and are rapidly oxidized by peroxides or permanganate to endosulfan sulphate with m.p. 181°C in acidic media the isomers hydrolyse to form endosulfan alcohol m.p. 203°C.

Endosulfan has a similar spectrum of insecticidal activity to aldrin except that it is also acaricidal.

❑❑❑

Chapter 24

Organophosphate Insecticides

Phosphamidon (Dimecron)

O, O-dimethyl-O (2-chloro-N, N-diethyl carbamyl)- Methylvinyl phosphate.

It is produced by reacting equimolecular quantities of trimethylphosphite with the boiling solution of diethyl amide of dichloroacetoacetic acid in chlorobenzene.

The yield of phosphamidon is about 83%. It is a colourless to pale yellow liquid. However the commercial formulation is bright violet in colour due to the presence of a dye. Its boiling point is 70°C. It has a faint but pleasant odour. It is highly soluble in water, alcohol, acetone and slightly soluble in saturated hydrocarbons such as hexane and practically insoluble in kerosene. Compared to other organophosphorus compounds, it has high water solubility and hence the presence of moisture in fillers and emulsifiers do not affect its stability. As this does not clog the nozzles, it is well suited for low volume and ultra low volume sprays

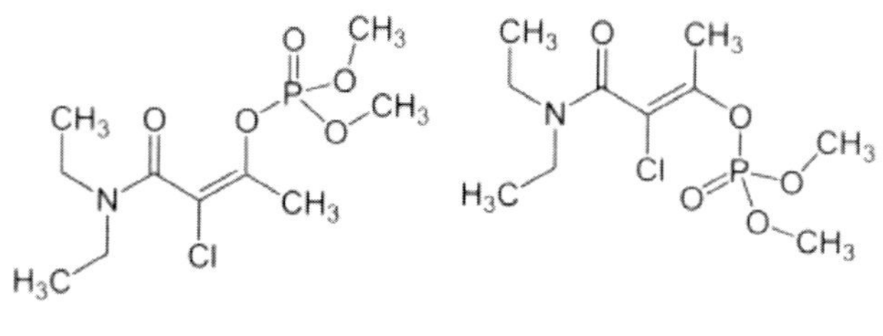

It is stable in neutral and weakly acidic aqueous solution. In alkaline medium it is rapidly hydrolysed. It is stable at least for two years when stored in closed containers. It is compatible with most pesticides except the alkaline ones like Bordeaux Mixture, lime sulphur wash, nicotine sulphate and oxychloride based fungicides. It gets absorbed inside the plant tissue within one to three hours and is translocated more towards the top. The active ingredients get degraded into less stable metabolites having low mammalian toxicity than the parent active ingredient. The insecticidal property is reduced when phosphamidon is mixed with copper oxychloride. It is non phytotoxic. When insects feed on the sap containing active material, the cholinesterase is inhibited and used for the sucking pests, leaf miners, certain mites etc.

Malathion (Carbophos)

(O, O-Dimethyl-S-1,2-dicarboethoxyethyl-dithiophosphate).

It is one of the safest organophosphorus compounds now marketed in the form of emulsion concentrate. It is also possible to use malathion in the form of dusts or suspensions prepared from a wettable powder.

It is produced by the addition of dimethyldithiophosphoric acid to maleic acid ester in the presence of basic catalysts in various organic solvents or without solvents.

The technical grade compound obtained by this method contains small amounts of impurities like trimethyldiophosphate, dimethylmaleate and solvent which decompose malathion rapidly. Xylene is most often used as a solvent for malathion synthesis. Decomposition takes place even at very low temperatures.

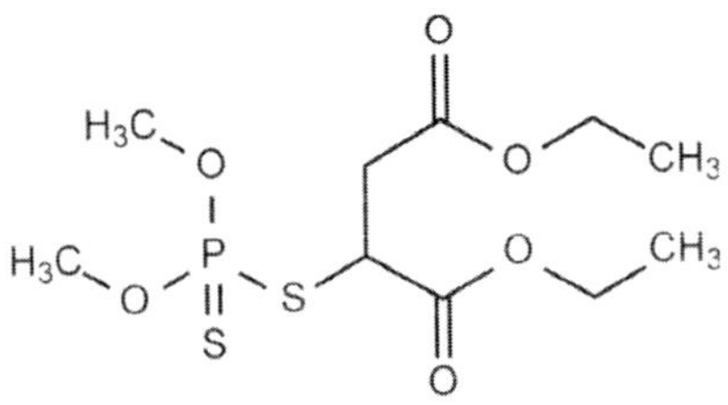

Pure malathion is a colourless liquid boiling at 120°C. At 20 °C, it is sparingly soluble in water but highly soluble in most organic solvents with the exception of saturated hydrocarbons. Malathion on prolonged heating at 150 °C is isomerised and goes over to the thiolo isomers. At higher temperatures the reaction proceeds violently and a considerable part of the product decomposes, sometimes even explosively.

In acid medium, the main products of hydrolysis are dimethyldithiophosphoric acid and the ester of mercaptosucinic acid while in alkaline medium and salt of dimethyldithiophosphoric acid and ester of fumaric acid are formed. When malathion is oxidized by nitric acid or by other strong oxidizing agents, the thion sulphur atom is split off and the corresponding ester of thiolo phosphoric acid is formed. Malathion on prolonged contact with iron or materials containing iron breaks down and completely loses its insecticidal properties. Therefore, it cannot be kept in iron containers. It is best kept in glass containers. It is non-phytotoxic. It is a non-systemic insecticide and acaricide of low mammalian toxicity useful in the control of a wide variety of crop pests especially in fruits, vegetables and stored products and external parasites of livestock.

Monocrotophos ($C_7 H_{14} O_5 NP$)

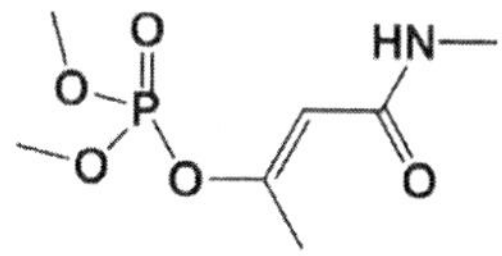

Dimethyl phosphate of 3 hydroxy-N-methyl-C is-crotonomide. It is a systemic insecticide and acaricide. Trade names are azodrin and Nuvacron. It is effective against thrips, leaf miners. It is effective against thrips, leaf miners, cotton bollworms and leaf webbers and certain mites on crops. It is toxic to bees.

It is manufactured by reaction of a dialkyl phosphorochloridate with a hydroxyl compound, generally in an organic solvent in presence of sodium carbonate or an organic base. The dialkyl phosphorochloridates can be easily manufactured from phosphorus oxychloride and an alcohol.

Chlorpyriphos ($C_9 H_{11} Cl_3 NO_3 PS$)

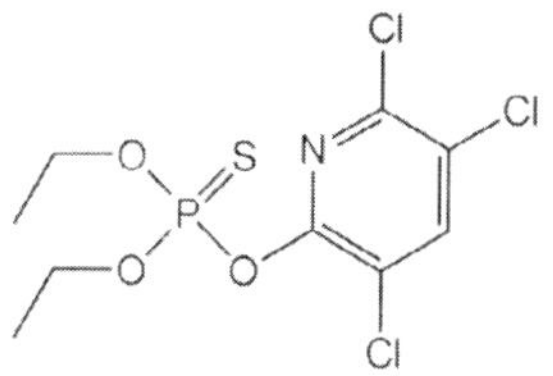

Trade name is Dursban. It is O, O-diethyl O-(3,5,6,trichloropyridyl-2) phosphorothinate. It is an active insecticide for the control of sucking and chewing plant pests, soil inhabiting plant pests and household parasites.
It is synthesized by reacting the sodium salt of 3, 5, 6, trichloro-2-hydroxy pyridine with diethyl chlorothiophosphate in dimethyl formamide.

It is a crystalline substance, M.P.41 to 43°C. It is highly soluble in most organic solvents, but almost insoluble in water. In acid or alkaline media, the compound is slowly hydrolysed by water forming diethylthiophosphoric acid and ethyl-thiophosphoric acids and trichlorohydroxypyridine. By oxidation it may be converted to the corresponding phosphate. The duration of action of the compound on different surfaces is 6 to 11 weeks, but on leaves of plants it is short acting apparently due to its rapid hydrolysis under the influence of enzymes. The compound is active on grain plants for several weeks.

Thimet (Phorate)

It is O, O-Diethyl-S-(ethylthiomethyl) dithiophosphate. It is a systemic insecticide of low water solubility applied to the soil for the protection of plants. It is produced by reacting diethyldithiophosphoric acid with formaldehyde and ethyl mercaptan at room temperature and then extracting the product with hydrocarbons or by reacting salts of diethyldithiophosphoric acid and chloromethyl ethylsulphide in ectonitrile or other suitable solvents. The content of the compound ranges from 70 to 80%.

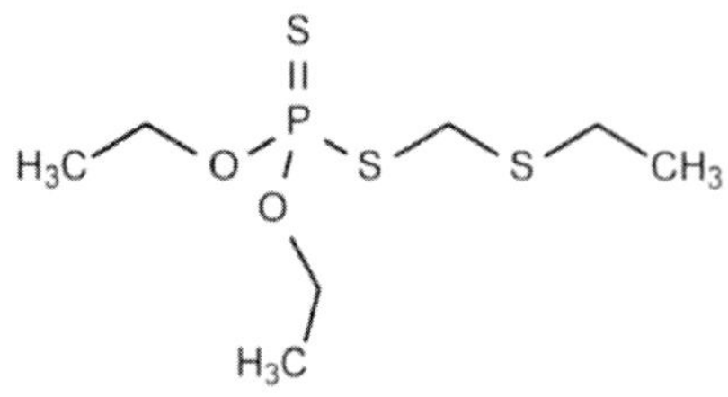

It is clear liquid, b.p. 100°C. It is highly soluble in most organic solvents, its solubility in water is about 70 mg/litre. It is relatively unstable to hydrolysis and at 70 °C and pH8, 50% of it is hydrolysed in two hours. In acid medium it is more stable. It is easily oxidized to the corresponding sulphoxide which is more resistant to hydrolysis than the starting compound. This explains the comparatively long insecticidal action of phorate in plants. The metabolites of phorate in most cases are as highly toxic to animals as the starting compound. The final products of metabolism are particularly biologically inactive biologically. It is mainly used for the treatment of seed at sowing. It has contact and fumigant action and to some extent nematicidal and acaricidal action also. It does not persist for a longer period and gets metaboligically oxidized yielding phosphorothioate and sulfone, which are readily hydrolysed. It has been found effective against sorghum shootfly and rice gallfly. Formulated as 10% granule.

Phosalone: ($C_{12}H_{15}ClNO_4PS_2$)

Trade name is zolone. It is O, O-Diethyl-S-(6-chlorobenzoxanol-methyl) dithiophosphate. It is produced by condensation of sodium of ammonium diethydiuthiophosphate with 6-chloro-3-chloromethyl-benzoxazolone.

It is practically insoluble in water but highly soluble in many organic solvents. This is a white crystalline substance with a garlic odour, m.p. 45 to 47°C. It is relatively stable in acid medium but in alkaline medium it is rapidly hydrolysed.

The main products of hydrolysis are 6-chloro-benezenehexazolone, diethylthiophosphoric acid and formaldehyde. It is a non-systemic contact insecticide and acaricide effective against a wide spectrum of pest species, particularly of cotton. It is safe to bees and natural enemies of pest species. Formulated at 4% dust and 35% EC.

Dimethoate (Rogor, Phosphamide, Bi-58)

Its chemical name is O, O-Dimethyl-S-(N-methyl carbanoylmethyl) dithiophosphate. It is a systemic and contact insecticide and acaricide. It has been recommended for the control of a wide variety of sucking pests, and lice infesting poultry. It is formulated as 30% EC. It is phytotoxic to certain varieties of sorghum, chrysanthemum, etc. It is proposed by reacting salts of dimethyl-dithiophosphoric acid with N-methylchloroacetamide, in aqueous medium in the presence of some organic solvents.

Pure dimethoate is a white crystalline substance with camphor-like odour. The technical material is a yellowish brown (amber) coloured oily liquid with sulphurous acid smell. The emulsion concentrate is a pale, straw coloured liquid with a characteristic pungent odour. Pure dimethoate melts at 51 to 52°C. It is highly soluble in paraffinic hydrocarbons. It is thermally unstable and on heating it decomposes. The O, S-dimethyl-S-(N-methyl carbonylmethyl) dithiophosphate obtained is more toxic to mammals than the starting dimethoate (LD_{50} for rats being 100 mg/kg). When it is oxidized by various oxidizing agents or by the oxygen of the air it is converted to O, O-dimethyl-S-(N-methyl carbanoylmethyl) thiophosphate. The hydrolysis of dimethoate takes place most rapidly in alkaline medium. In acid medium it is more stable. Depending on the dosage of dimethoate applied, it is usually broken down in plants in 15 to 20 days.

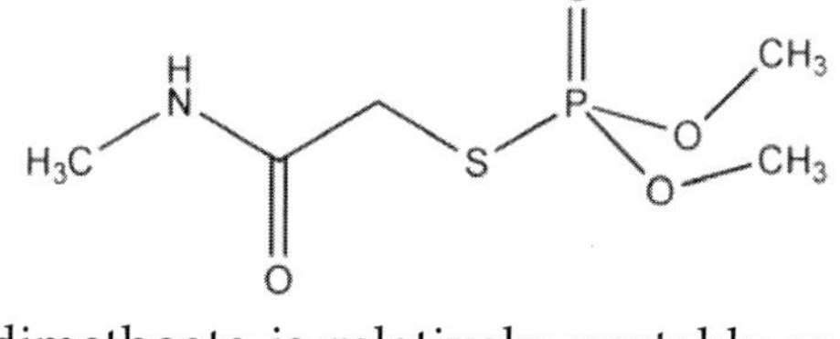

In storage, dimethoate is relatively unstable and rapidly breaks down particularly at an elevated temperature. Impurities present in dimethoate catalyse its decomposition. It decomposes very quickly when it contains traces of organic bases or comes in contact with iron or salts of iron. Hence iron containers should not be used for storing dimethoate. Solutions of dimethoate in organic solvents are more stable. It is marketed in the form of 40% emulsion concentrate. The LD_{50} of pure compound is 250 to 265 mg/kg while that of the technical material is 150 to 165 mg/kg.

Quinalphos (Ekalux) C_{12} H_{15} O_3 N_2 PS

It is O, O-diethyl-O-quinoxalinyl-2-thionophosphate. It is formulated as 25% EC, 5% granule and 1.5% dust. It is effective against sucking insects and lepidopterous larvae particularly of cotton and rice. It is also an effective acaricide. LD_{50} for rat; oral 62-137 mg/kg.

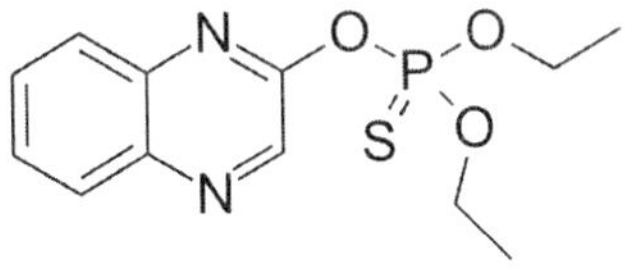

□□□

Chapter 25

Carbamate Insecticides

Carbamates are derivatives (esters) of carbamic acid. Carbamic acid (Sevin, Naphthyl carbamate) $C_{12}H_{11}NO_2$. Its chemical name is 1-Naphthyl-N-methyl carbonate. It is a good contact insecticide with occasional systemic action in certain cases. It is produced by reacting 1-naphthol with methyl carbanyl chloride at room temperature for about 10 to 20 hours.

The rate of reaction is increased by removing the HCl that is formed with a stream of air or nitrogen. Purer compound of carbaryl is obtained by reacting 1-naphthol with methylisocyanate. It is also synthesized by reacting 1-naphthyl-chlorocarbomate with methylamine in the presence of HCl acceptors. Naphthyl-N-methylcarbamate is a white crystalline compound, m.p.142 °C . It is soluble to less than 0.1% in water but highly soluble in organic solvents. At room temperatures, it is resistant to the action fo water, light and oxygen of the air. In alkaline medium it is rapidly hydrolysed and therefore it is not compatible with compounds of alkaline nature like Bordeaux mixture, calcium and barium polysulphides.

The technical material is crystalline and is unstable. It is only weakly phytotoxic. It has been established that the principal metabolites of decomposition of carbaryl in plants and animals are products of oxidation and hydrolysis. Oxidation is directed primarily at the methyl group attached to the nitrogen atom. The other metabolites 4-hydroxy and 5-hydroxynaphthyl-N-methylcarbamates that are formed are further broken down with rupture of the aromatic ring. It is marketed in the form of dusts and wettable powders. The LD_{50} for rats is 500 mg/kg. The compound does not irritate human skin. It is compatible with most pesticides excepting Bordeaux mixture, lime sulphur and urea. It is effective against a wide spectrum of insect pests of crops particularly

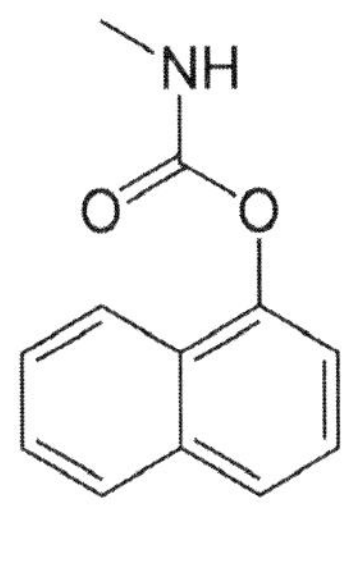

of cotton. It does not kill mites. Carbaryl dust finds use in the control of human lice and dandruff and cattle lice.

Carbofuran (Furadan) $C_{12}H_{15}NO_3$

Its chemical name is 2,3 Dihydro-2,2-dimethyl-7-benzofuranyl methylcarbamate). It is a systemic carbamate insecticide of broad spectrum utility. It is white crystalline substance, odourless when pure, m.p.153-154 °C. It is stable in acid and neutral media, but unstable in alkaline medium. It is sparingly soluble in water, slightly soluble in organic solvents, methylene chloride. It is compatible with all non-alkaline pesticides and fertilizers. It is not phytotoxic to rice: Radioactive carbon studies have shown that carbofuran is very rapidly taken into roots and distributed into the aerial parts of plants. The oral LD_{50} for rats is 11 mg/kg. It is a systemic insecticide and nematicide effective against sucking and soil inhabiting pests and in the control of sorghum shootfly.

It is made by reaction of the hydroxyhecterocyclic compound with methyl isocyanate. It is broad spectrum insecticide, nematicide and miticide.

LD_{50} : 8-14 mg/kg rat

It is stable in acid and neutral media but unstable in alkaline medium. Sparingly soluble in water; but soluble in organic solvents. It is compatible with non-alkaline pesticides and fertilizers. It is not phytotoxic to rice.

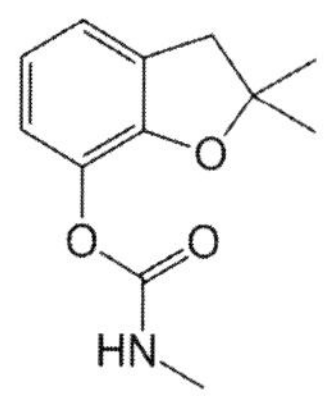

Aldicarb (Temik) $C_7H_{14}N_2O_3S$

Its chemical name is 2-Methyl-2-methylthiopropyonal-oxime-O, N-methyl carbamate. It is a systemic insecticide, acaricide and nematicide. It is synthesized by reacting the corresponding oxime with methylisocyanate.

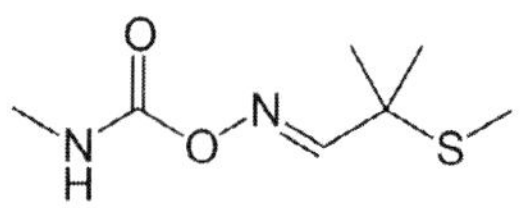

It is a white crystalline substance with a m.p. 100 °C. It is soluble in paraffinic hydrocarbons, sparingly soluble in water, soluble in acetone, ethyl alcohol, chloroform and chloro benzene. Its LD_{50} for rats is 0.93 mg/kg. It is effective against insects, mites and nematodes.

Carbosulfan

Carbosulfan, a sulphenylated derivative of carbofuran acts as a contact and systemic insecticide. It can be applied to the foliage or soil as a nematicide. It has a lower mammalian toxicity.

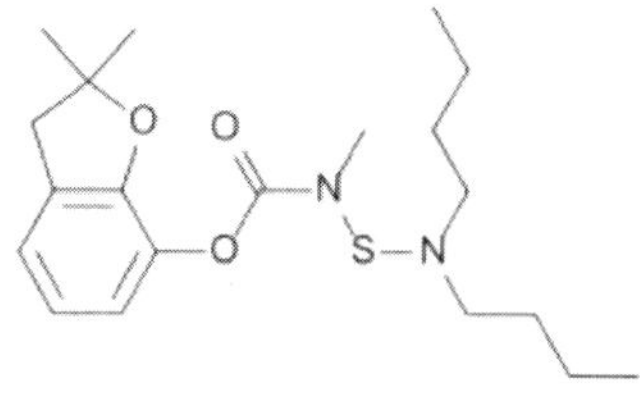

The carbamates, carbofuran, carbosulfan and aldicarb are valuable nematicides. Carbofuran, a systemic nematicide is used for corn, sugar beet, peanut, sugarcane, tobacco and vegetable. It has a comparatively short residual life. Carbosulfan formulated as granules is used in vegetables, eg. brassicas, carrots and turnips.

Aldicarb is highly active systemic nematicide but is highly toxic to man and animals. It is formulated as granules for soil application. Aldicarb is used for a wide variety of crops like vegetables, strawberries and ornamentals.

The organophosphorus and carbamate nematicides toxicity is due to poisoning of the enzyme cholinesterate by phosphorylation or carbamoylation.

□□□

Chapter 26

Botanicals Insecticides

Toxicants derived from plants are used in insect control and among them nicotine and pyrethrum are well known.

Nicotine ($C_{10} H_{14} N_2$)

Its chemical name is 1-methyl-2,3 (pyridyl pyrrolidine). It is the principal alkaloid in tobacco (An alkaloid may be defined in general as a naturally occurring heterocyclic, optically active nitrogenous base of relatively high molecular weight and having marked physiological activity). Nicotine is found in the leaves of *Nicotiana tabacum* and *Nicotiana rustica* from 2 to 14 per cent. Among the twelve alkaloids present in tobacco nicotine is the most important one contributing about 97 per cent and the other two of insecticidal value are (i) Nornicotine ($C_9 H_{12} N_2$) 2-(3-pyridyl pyrrolidine), (ii) anabasine (Neonicotine, 3-(2-piperdyl) pyridine.

Nicotine is present in all stages and in all parts of the tobacco plant, the quantity ranging from a trace in seedlings to a maximum in mature leaves. It is extracted from the woody portions of the plant such as stems and leaf midribs unsuited for smoking or chewing tobacco. It is extracted with an organic solvent such as benezene, a mixture of tobacco refuse, milk of lime and caustic soda. The extract is concentrated in vacuum and again extracted with dilute H_2SO_4, the nicotine then liberated with alkali and extracted with a mixture of ether and petroleum ether. Subsequently by distillation, pure nicotine is obtained.

Pure nicotine is a colourless, odourless liquid, the alkaloid boiling at 247.3 °C with a specific gravity of 1.0093. It is soluble in water in all proportions only below 60 °C and above 210 °C where it forms hydrates. Under room temperature it is soluble in organic solvents. When exposed to air and light, it slowly darkens in colour and becomes more viscous. Nicotine as a free base was found to be five to seven times as toxic as nicotine sulphate. The efficiency of nicotine sulphate sprays increases as the alkalinity of the spray water is increased. Nicotine sulphate containing 40% alkaloids is safer, more convenient to use and the free alkaloid is liberated by the addition of soap, lime or NH_4OH to the spray solution. It was used for the control of cardamom thrips. Dust formulation of nicotine sulphate releases nicotine in the presence of moisture.

Tobacco decoction, useful for controlling aphids, thrips, etc. is prepared as follows. Boil 1 kg of tobacco waste in 10 litres of water for 30 minutes or keep it in cold water for a day. Then make up the infusion to 30 litres and add about 90g of soap. Addition of soap improves wetting, spreading and killing properties.

Insecticidal action of nicotine

It is a nerve poison being highly toxic when absorbed through the cuticle or taken in through the tracheae or when ingested. Nicotine and anabasine affect the ganalion, blocking conduction at higher levels. The fumes of nicotine, striking the walls of insects are immediately dissolved, penetrated to vital tissues and bring paralysis of the insects nervous system.

Nornicotine

It is similar to nicotine except that the methyl group attached to pyrrolidine ring has been replaced by a hydrogen atom. It occurs in *Nicotiana sylvestris*, a species of tobacco. In this plant, 95% of the alkaloid present in nornicotine. It is a liquid alkaloid resembling nicotine in physical properties and physiological behaviour. Pure nornicotine is a colourless, hygroscopic, slightly viscous liquid with an odourless pungent than nicotine. It is slightly volatile in steam. The specific gravity is slightly greater than nicotine. It was found to be more poisonous than nicotine.

Anabasine (Neonicotine)

It occurs in seeds of *Anabasis aphylla*. It closely resembles nicotine in its physical and chemical properties. It is a colourless, some what viscous liquid which turns brown on standing in contact with air. It is very stable and possesses an odour similar to that of nicotine. It is miscible with water in all proportions and is soluble in all organic solvents. It is extracted from its plant source by ethylene dichloride or by a water kerosene diffusion process. The crude anabasine sulphate of commerce is a mixture of alkaloids (i) aphillidine ($C_{15}H_{22}ON_2$), aphylline ($C_{15}H_{24}ON_2$) and lupenine ($C_{10}H_{19}ON$) and methyl anabasine. It has been found to be approximately equal to nicotine in its insecticidal activity and toxicity to warm blooded animals.

Nicotine and to lesser extent anabasine are used as contact insecticide for aphids attacking fruits, vegetables and ornamentals and as fumigants in green house and against poultry mites. Nicotine's rapid deterioration in light and air favours its use on leafy vegetables within a few days of harvest. Nicotine sulphate is safer and more convenient to use.

Pyrethrum

The insecticidal principle in pyrethrum is found in the flower heads of certain plants of Chrysanthemum genus, family compositeae. Only a few species like *Chrysanthemum roseum*, *Chrysanthemum cineravieflium*, *Chrysanthemum marshalli* and *Chrysanthemum tamrutene* have been found to be valuable sources of this insecticide. It is used as dusts and sprays.

The flowers are dried at 54.4 °C. These flowers are ground to a fine powder and extracted with the solvents. The solvent is repeatedly percolated through the ground flowers and the weak extracts are evaporated in a vacuum still to recover the solvent. The extracted powder (pyrethrum mavc) containing traces of pyrethrins is sometimes used as carriers in dust preparations. Dust concentrates are made from concentrated pyrethrum extracts with a non-volatile solvent and a suitable absorbent carrier and then diluted before use with the inert diluent. Antioxidants such as tannic acid or hydroquinone is used to stabilize the pyrethrins in dust preparations. Concentrated extracts of pyrethrum in acetone, alcohol or a hydrocarbon solvent together with an emulsifier is sold in the market.

The structure of the active principle in pyrethrum was established in 1950's. Standinger and Ruzicka found two active principles Pyrethrin I and Pyrethrin II which were esters of a Keto alcohol (Pyrethrolone) and two acids (Chrysanthamam monocarboxylic acid and Chrysanthamam dicarboxylic acid). Two more compounds were identified. These compounds are all viscous liquids, soluble in a variety of solvents but not in water. The active principles of a typical extract of pyrethrum is

Pyrethrin I - 35%
Pyrethrin II - 32%
Cinerin I - 10%

Cinerin II	-	14%
Jasmolin I	-	5%
Jasmolin II	-	4%

Since they are esters they are easily decomposed in the presence of moisture and acid or alkalines. Because of this, it is not compatible with lime or soaps. Cinerins are more stable than pyrethrins. As these compounds are not volatile, it has no fumigant value. The insecticidal properties of the pyrethrum are due to five esters (Pyrethrins I and II, Cinerins I and II and Jasmolin II) that are present mostly in the acheness of the flowers, ranging from 0.7 to 3.0%.

Insecticidal activity

Pyrethrins are more toxic than cinerins and variations in their activity depends on the insect species and the technique employed in any particular test. Pyrethrum affects the nervous system of insects such as cock roaches, resulting in muscular excitation, convulsion and paralysis. The affected cockroaches emerge from their hiding places, fly and become paralysed. The solubility of pyrethrum in oils promotes their penetrations of the insect integuments. The finer the pyrethrum powder the more rapid is its action on insects. The most active isomers are the a trans acid esters of the a cis alcohols. They are of low mammalian toxicity because of their detoxification following oral ingestion. Oral LD_{50} value 820 mg/kg.

Uses

In aerosol spray for indoor use against flies and mosquitoes. Kenya, is the main example in producing both coil and aerosol insecticides based on pyrethrum.

Pyrethrum Synergists

An examination of the constituents of sesame oil, which had been shown to increase the activity of pyrethrum, led to the development of a number of compounds containing the methylene-dioxyphenyl group as synergists. The most widely synergist for the pyrethrins is an ether of prophylpiperonyl alcohol always known by the pseudo systematic name of piperonyl butoxide.

❑❑❑

Chapter 27

Synthetic Pyrethroids

Fenvalerate

At Rothamsted Experimental Station in 1972 an exceptionally valuable combination of properties was found in the esters of 3-phenoxybenzyl alcohol with Cis-and trans-dichlorovinyl analogues of chrysanthemic acid having chlorine atom in place of the methyl groups in the isobutenyl side chain of which permethrin is an example.

The further important discovery that the cyclopropane ring could be replaced by an open chain came from Japan in 1974 when Ohno and coworkers showed that highly active esters (e.g. S-5439) could be obtained when various 2-aryl-3-methylbutyric acids replaced chrysanthemic acid. Further increase of insecticidal potency was generally obtained by substituting and a-cyano group into 3-phenoxybenzyl esters (e.g. fenvalerate).

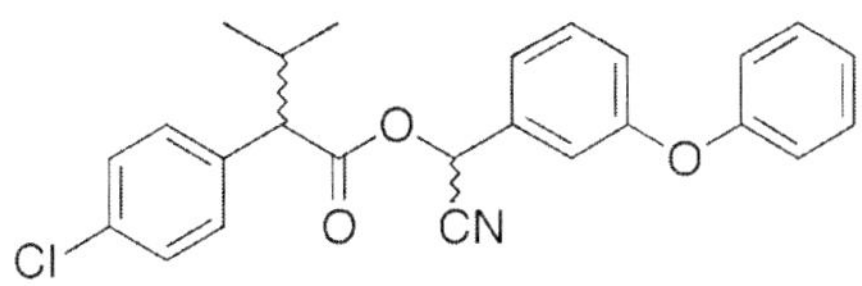

It is a contact and stomach insecticide for a wide range of insects that are resistant to organochlorine ,organophosphate and carbomate insecticides. It control flies, bollworms and other sucking insects.

Cypermethrin

The ester formed between a-cyano-3-phenoxybenzyl alcohol and (IR, Cis)-3-(2,2-dichlorovincyl)-2,2-dimethylcyclopropane carboxylic acid gave an active compound from which a single isomer cypermethrin was separated.

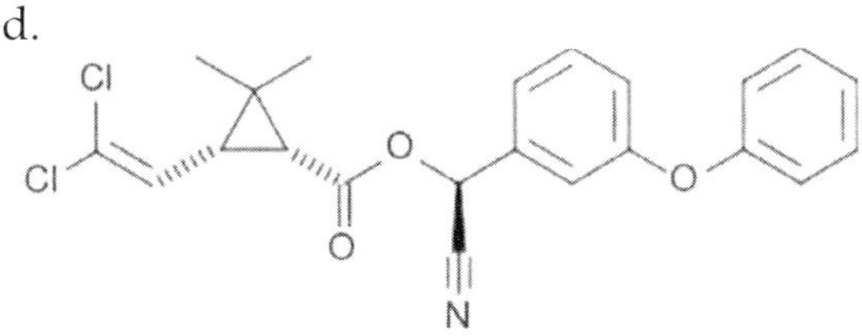

Fluvalinate

Developments in the 3-methylbutyrate series include fluvalinate.

The synthetic pyrethroids are very expensive to prepare on a tonnage basis. The high insecticidal activity and low mammalian toxicity of pyrethroids are especially significant now that compounds stable to light and oxygen are potentially available. Their toxicity to fish is high. They are rapidly degraded in soil and have no detectable ill-effects on soil microflora and microfauna. They are not active against mites.

The major symptoms of pyrethroid poisoning in insects may be accounted for by effects on the kinetics of nerve membrane sodium channels. The mean open times of these channels are prolonged with consequent hyperactivity of nerves.

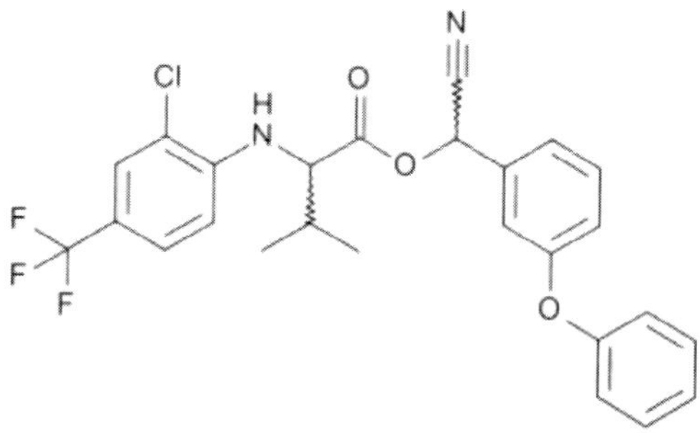

The synthetic pyrethroids have been found to be useful as early season sprays to control the variety of insects that occur on cotton including boll worms, leafworms, jassids, thrips and whitefly. They are used in combination with an organo phosphorus insecticide as an ultra-low volume spray. They are non toxic to humans and animals. LD_{50} to rats is around 8000 mg/kg. They are used at only 50 g/ha. The cost of treatment per hectare is low.

□□□

Chapter 28

Chemistry of Fungicides

In India, Lawrence in 1904 used the Bordeaux mixture for the first time. Subsequently a number of synthetic fungicides were developed and have been widely used in India.

Classification

1. Sulphur fungicides
   - (i) Inorganic
   - (ii) Organic
2. Copper fungicides
3. Mercury fungicides
   - (i) Inorganic
   - (ii) Organic
4. Others
   - (i) Quinones
   - (ii) Heterocyclic N compounds
   - (iii) Organo tin compounds
   - (iv) Antibiotics
   - (v) Systemic fungicides

Sulphur compounds

It is one of the most important fungicides used for the control of various diseases, particularly powdery mildews. Elemental sulphur occurs in both free state and also in combination. It occurs in many allotrophic forms, of which rhombic form, is the stable one at ordinary temperatures. It is insoluble sulphur chloride and benzene. Sulphur is widely used in a finely ground condition or in the colloidal form in dust preparations. Finely ground sulphur is obtained by grinding coarse sulphur in special types of mills called micronizers and then fractionating it with the air blasts. As the mechanically ground sulphur tends to aggregate, small amounts of inert material like kaolin or lead arsenate are also added to the micronised sulphur. The fungicidal efficiency of sulphur dusts deopends on the fineness of the particles.

Wettable powders are made by grinding sulphur with protective colloidal materials such as casein, bentonite clay, skimmed milk, flour or dextrin. Sulphur dusting powders containing materials such as gypsum, bentonite or talc in small amounts make the sulphur flow freely and this form is called "conditioned sulphur".

Hydrophylic oil is prepared by passing H_2S through a saturated solution of SO_2 in water or by reaction of $Na_2 S_2 O_4$ and $H_2 SO_4$ or mineral acids and a hydrophobic colloidal sulphur prepared by acidifying lime-sulphur solution with mineral acids. The hydrophilic sulphur preparations were found to have greater toxicity than hydrophobic suspensions, probably due to differences in the particle size.

Another form of sulphur called "floatation sulphur" is a by product of the manufacture of fuel gas from coal. It is marketed both as a paste and as wettable powders. In addition to the free sulphur, number of sulphur compounds have also been tried as fungicides. Among them lime sulphur or calcium polysulphide is an important one. Lime sulphur is prepared by boiling 1.8 parts of sulphur, 1 part of lime and 10 parts of water for one hour. Then the solution is filtered and the filtered solution is diluted with water to specific gravity of 1.003 and 1.035 for summer and dormant sprays respectively.

Fungicidal action of sulphur

Sulphur acts as a hydrogen acceptor in the metabolic system and by doing this disturbs the normal hydrogenation and dehydrogenation reactions in the cells. It acts much like oxygen and forms H_2S instead of H_2O in the fermentation process thus causing a disturbance in the kerb's cycle. Sulphur fungicides emit sufficient vapour to present the growth of fungus spores at a distance of several millimeters from deposits on leaves.

Copper compounds

Bordeaux Mixture

A 4-4-50 mixture (copper sulphate in lb: lime in lb: water in gallons) is used but the concentration of the ingredients is varied, depending on the purpose. In India Bordeaux mixture is being made by preparing a solution of copper sulphate and quick lime (or hydrated lime) in finely ground form in separate containers and then mixing them simultaneously into a third container with constant agitation.

Freshly made Bordeaux mixture is of gelatinous nature. The colour is deep blue to slightly greenish. The presence of $CaCO_3$ in the lime hastens the rate of settling and reduces the volume of the precipitate. Therefore, the proportion of $CaCO_3$ should not be more than 25%. After preparing the Bordeaux mixture, it should be tested for free copper as it is phytotoxic. This is done by adding a few drops of 10% solution of potassium ferrocyanide to a small quantity of the prepared Bordeaux mixture. If a reddish brown deposit is formed immediately, more lime should be added to the mixture. If potassium ferrocyanide is not available, a clear iron blade could be dipped inside the mixture for sometime and then the presence of copper deposit observed.

Bordeaux mixture should be sprayed while fresh. On standing, it looses its fungicidal property. However the mixture may be stabilized and kept for several days by adding sugar or jaggery at the rate of ½ lb in 50 gallons of mixture.

Mechanism of action

Accumulation of copper is the mechanism by which copper from Bordeaux mixture becomes toxic to the spores. The fungicidal action of Bordeaux mixture is more protective than eradicative. Due to the presence of an excess of $Ca(OH)_2$, it cannot be combined with soaps or pyrethrum which are decomposed by free alkalies. It is compatible with nicotine lead and calcium arsenate and oil emulsion. It is a fairly efficient emulsifier. It exhibits phytotoxicity on crops like apples and peaches.

Copper oxychlorides (Blue Copper 50 : Fytolan)

It is one of the low soluble copper fungicides produced by the action of air on cupric chloride soluble or scrap copper. It is marketed in the form of wettable powder containing 50 and 90% copper oxychloride and dusts containing 4 to 12% metallic copper. The 50% formulation contains a diluent, usually kaolin and a surface active agent.

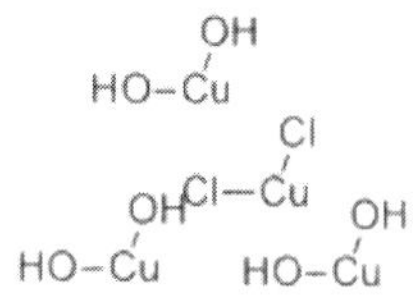

Compared to the Bordeaux mixture, copper phosphate and basic copper sulphate are more inferior in toxicity while copper oxychloride, copper oxalate and copper zeolite are intermediate. Among the various copper fungicides tested, no copper compound has high fungicidal value with low phytocidal properties. Bordeaux mixture is the best possible compromise available now.

❑❑❑

Chapter 29

Organic Fungicides

Organic sulphur compounds-Dithiocarbamates

The derivatives of dithiocarbamic acid had insecticidal and fungicidal properties.

I. Dithiocarbamates (organic S compounds)
II. Thiuram disulphate - Thiram
III. Bis dithio carbomates

Zineb (Dithane Z-78, Parazate)

It is one of the most important fungicides used in agriculture. It is prepared by precipitation of the zinc ethylene bisdithiocarbamate from aqueous solution of diammonium or disodium ethylene bisdithiocarbamates by $ZnSO_4$.

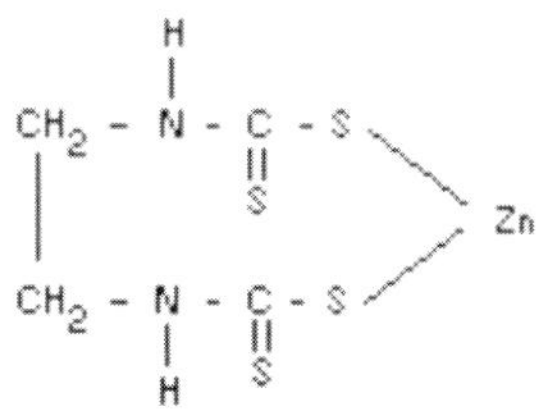

This reaction is carried out in dilute solution (<7%) as polymeric zineb of lower toxicity precipitates from concentrated solutions or by reacting ethylene diamide with zinc oxide and carbon disulphide. It is a white crystalline substance with molecular weight 275.5, almost insoluble in water and in most organic solvents but moderately soluble in pyridine. It is unstable in the presence of heat, light and moisture. The moist compound (>4%) under unfavourable storage conditions may decompose even up to 50% in a year. To reduce the changes of explosive evolution of C S_2, zineb has to be stored in well ventilated places at the lowest possible temperatures.

It breaks down rapidly (10 to 15 days) on plants with the final formation of easily volatile products. The breakdown takes place considerably faster in warmer and clearer weather. The final product is $ZnSO_4$ which may play the role of Zn micronutrient. It can be combined with parathion, metasystox, systox and wettable sulphur. In the presence of strong alkaline substances it may be decompose. It may cause irritation of nose, throat and skin. It is formulated as wettable powders. It gives better control of late blight of potato.

Maneb (Dithane M-45, Manzate)

Maneb, the manganese salt of ethylene bisdithiocarbamic acid, is similar to zineb in most of its physical and chemical properties. It is produced by reacting manganese sulphate with diammonium or disodium ethylene bisodithiocarbamate in aqueous solution.

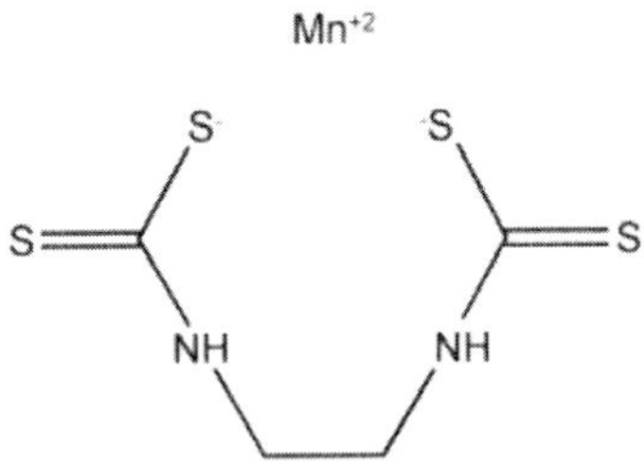

It is a yellow crystalline. Substances with a molecular wt. of 265.3, it is insoluble in water and organic solvents. The pure compound breaks down at about 120°C. When stored in bulk, it is capable of spontaneous decomposition with charring. However this hazard is reduced when the compound is diluted with an inert diluent. Regarding compatibility, it is similar to ferbam.

Mode of action

According to Sisler, the toxicity of dialkyldithiocarbamates in the cell might involve one or more of the following.

- Chelation of required heavy metals.
- Attachment of a 1:1 complex metal and dithiocarbamate ions to enzyme.
- Attachment of dithiocarbamate ion to metals bound to protein.
- Reaction of free radical intermediate with cellular compounds.
- Lethal catalysis as suggested by Lowe and Phillips.

Inspite of all the above possible mechanisms the actual mode of action of dialkyldithiocarbamate in fungus cells remain obscure. Increasing the size of the alkyl groups on the nitrogen atom of dialkyldithiocarbamates greatly decreases the toxicity.

❑❑❑

Chapter 30

Systemic Fungicides

The idea of controlling plant diseases by internal treatment of plants is only since 1960s that commercially viable systemic fungicides have come into the market.

A systemic fungicide is a compound that is taken up by a plant and is then translocated within the plant protecting it from attack by pathogenic fungi or limiting an established infection.

A systemic fungicide should possess the following properties.

- It must be fungicidal or converted into an active fungi toxicant on the host plant.
- It must possess very low phytotoxicity.
- It must be capable of being adsorbed by roots, seeds or leaves and translocated.

The earlier protectant or surface fungicides, applied as foliar sprays, formed by deposits protecting the plant from fungal attack. The deposits may be removed in course of time and also cannot protect the new plant growth. Hence the use of systemic fungicides has come into practice.

There are now a number of systemic fungicides on the market which may be divided into the following eight groups.

Benzimidazoles,
Thiophanates,
Carboxamides,
Pyrimidines,
Pyridines,
Morpholines,
Azoles and
Organophosphorus compounds

Benzimidazoles

These represented a new era in fungicide use when they were introduced in late 1960s. The most important group members are:

Benomyl (Benlate) (methyl-l-(butylcarbamoyl) benzimidazole-2-carbamate).

Thiabendazole (2-4-thiazolyl) benzimidazole.

Benomyl

Benomyl was introduced in 1967. It was synthesized from cyanide and methyl-chloroformate. Benomyl and thiabendazole are both wide spectrum systemic fungicides active against many pathogenic fungi including mildews and soil-borne pathogens, *Verticillium alboatrum* on cotton and black spot on roses.

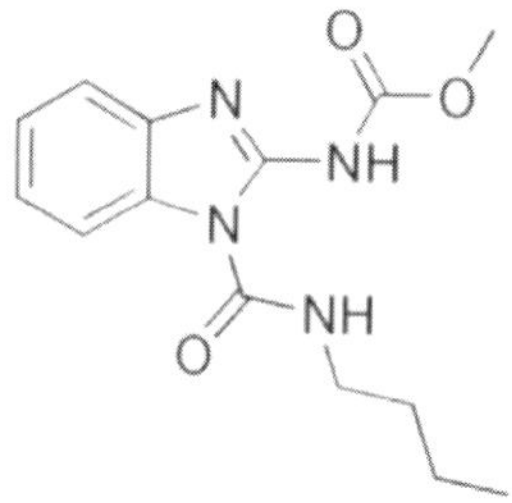

Structure of Benomyl

Du-Pont; Benlate LD_{50} > 9590 mg kg^{-1} rat

Foliar fungicide in wine grape fruits, vegetables, citrus, cereal seed dressing.

Benomyl is the more active compound and is widely applied as a foliar spray, seed dressing or to the soil for control of grey mold (*Botrytis cinerea*), apple scab (*Venturia inequalis*) canker and powdery mildew (*Podosphaera leucotricha*), leaf spot (*Cercospora beticola*), major fungal diseases of soft fruits and some pathogens of tomato and cucumber.

Thiabendazole (Mertect) is good against post-harvest diseases of apples and pears and as a seed dressing in bunt disease of wheat. It is also used against stem rot of bananas.

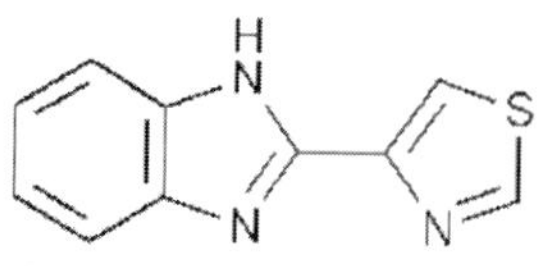

Thiabendazole X = S

Fuberidazole X = O

Fuberidazole (Voronit) is a valuable seed dressing chemical against Fusarium diseases (eg. of rye and peas), wheat rust (*Puccinia triticina*) and barley powdery mildew.

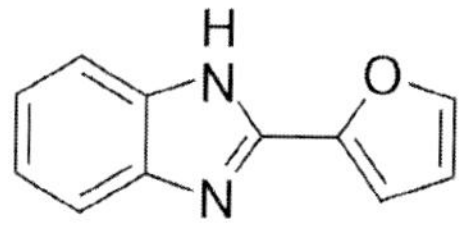

The fungicidal activity (fungitoxicity) is associated with the benzimidazole nucleus. In aqueous solution benomyl is rapidly hydrolysed to the strongly fungicidal methylbenzimidazole-2-carbamate and this is probably the active fungitoxicant.

Carbendazim (Bavistin, Derosol)

Foliar fungicide in grapes, fruits, vegetables, cereals, cereal seed dressing LD_{50} : 15,000 mg kg^{-1} rat.

Carbendazim (Methylbenimidazole-2-Carbamate) is used as a wide spectrum systemic fungicide and may be formulated as a 50% w.p. for control of Botrytis, Gloeosporium rots, powdery mildews and apple scab. Carbendazim is absorbed by the roots and foliage of plants and is quicker acting than Benomyl.

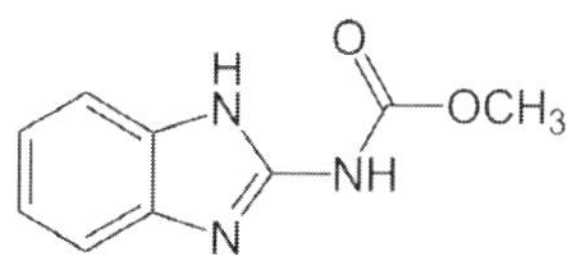

The activity of the benzimidazole fungicides (Benomyl, thiabendazole and carbendazim) is due to the inhibition of nuclear division due to their action on the microtubule assembly and the resistance developed in fungi is the result of mutant strains possessing an altered microtubule assemble.

Carboxin and Oxycarboxin

Oxathins are another group of heterocyclic compounds with systemic fungicidal properties. Examples are Carboxin (5,6 dihydro-2-methyl-l, 4, oxathin-3-carboxanilide) (vitavaz) and the sulphone analogue known as Oxycarboxin (Plantvax).

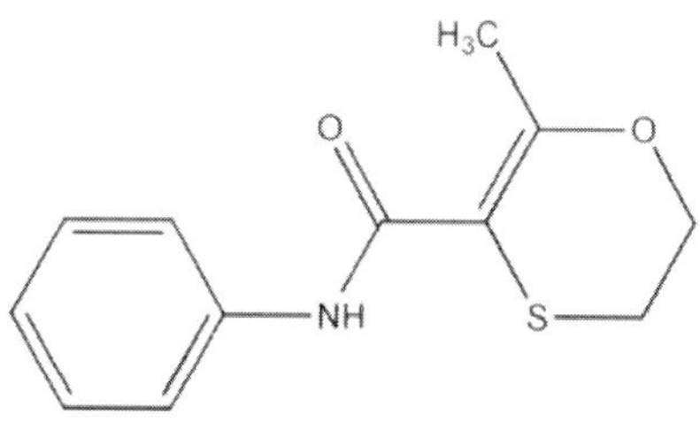

Seed dressing: Cereals, cotton, soil and foliar fungicide.

Smuts and Rusts

Carboxin is prepared by reaction of a-chloroacetoacetanilide and 2-thiothanol followed by cyclization.

Oxycarboxin is obtained by subsequent oxidation of carboxin with hydrogen peroxide. Both are fairly water soluble and are not phytotoxic.

They are active against Basiodiomycetes class of fungi causing rusts, smuts and bunts of cereals and soil fungi *Rhizoctonia solani*. Carboxin can be formulated with other fungicides like thiram, copper oxine.

Oxycarboxin

Oxycarboxin has systemic action against rusts of cereals, and vegetables and seed treatment or soil application can be done.

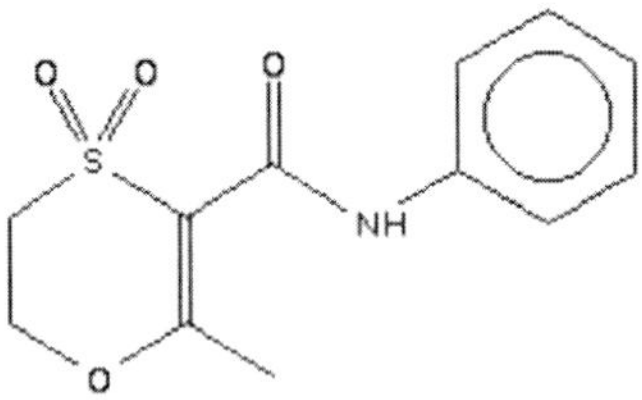

Carboxin is absorbed and translocated by plant roots. In water, soil and plants; the compound is oxidized to sulphoxide but further oxidation to sulphone was not observed. The sulphoxide is much less fungicidal and so oxidation causes loss of activity.

The primary mode of action of carboxin and related compounds probably involves the blocking of succinate oxidation in the mitochondria of sensitive fungi.

Metalaxyl

Metalaxyl is included in the phenylamide group of systemic fungicides.

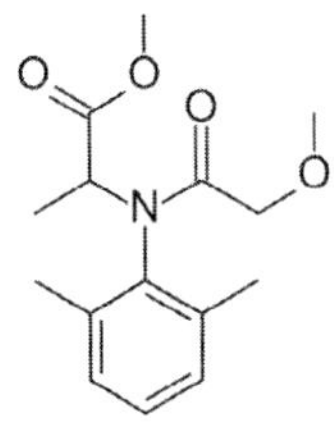

The compounds of this group show protective and systemic activity against Oomycetes causing foliar, root and crown diseases in wide range of crops eg. Downy mildews and late blight.

The first members of this group Metalaxyl and Furalaxyl were introduced by Ciba-Geigy in 1977 (Metalaxyl = Ridomil).

Metalaxyl LD_{50} L 669 mg kg^{-1} rat

With a high activity at low rates of foliar or soil application metalaxyl controls diseases caused by air or soil borne oomycetes in crops like potatoes, grapes, tobacco, cereals and vegetables.

A wettable powder formulation with mancozeb, (a complex of Zn and Mn salts fungicide) is widely used as foliar spray against blight on potatoes. Metalaxyl has the broadest spectrum of fungicidal activity of this group of fungicides; it is good against downy mildew on vines, lettuce, maize and Pythium diseases and can be formulated as a seed dressing.

❑❑❑

Chapter 31

Chemistry of Herbicides

A herbicide in the broadest sense is any compound that is capable of killing or severely injuring plants and may be used for elimination of plant growth.

A weed is any plant, wild or cultivated that is undesired in that particular place. In agriculture and horticulture weeds are thus any plant other than the specific crop being grown. On railway tracks, industrial sites, air port paths, open spaces and the like the entire vegetation can be regarded as weeds.

Weeds are conveniently divided into dicotyledonous plants, termed broadleaf weeds and monocotyledonous plants, termed grass weeds. Weeds compete with plants for water, light, food and above and below the surface. The yield loss due to weeds is estimated to be 9-10%. Also herbicide economies or helps in crop production by reducing the cultural operations.

Herbicide may be classified based on one or more common characteristics such as chemical composition or mode of action or time of application etc.

Chemical	(i)	Organic
	(ii)	Inorganic
Mode of Action	(i)	Contact
	(ii)	Systemic
Time of application	(i)	Pre-Sowing
	(ii)	Pre-emergence
	(iii)	Post-emergence of crops
	(iv)	Pre-emergence or post-emergence is with respect to the emergence of crop and not of the weed

Manner of adsorption	(i)	Soil herbicides-root absorptions
	(ii)	Systemic
	(iii)	Foliage herbicide-contact
Range of Application	(i)	Total-unspecified vegetation
	(ii)	Selective

The type of herbicide used is selected according to the nature and state of development of the target weed. Soil-applied herbicides are used for weeds not yet emerged; foliar or soil applied herbicides are used as post-emergence.

The largest and economically most important group of agents used today is that of the selective herbicides which find application in all crops. A selective herbicide has high activity against the target weed but is well tolerated by the crop plant.

It is not possible to control all the target weed species simultaneously with one agent and hence various combinations are used in practice to achieve control. According to the type of problem, typical soil-applied herbicides (to prevent new growth) foliar applied agents (to kill of emerged weeds) or mixtures of both are used.

Selectivity of the Herbicidal Action

The causes for the difference in sensitivity of the plants to herbicides are very complex. The relationship between structure, physico chemical properties and selective herbicidal action of a chemical are all important factors. Complex interaction of properties of the chemical, the plant contacted with it (morphology, physiology, biochemistry) and its environment (soil factors, humidity, light, temperature) is responsible for the selectivity of herbicides.

The causes of herbicidal selectivity may be of two groups, viz., physical and biochemical.

- Physical causes of herbicidal selectivity include application techniques that lead to a spatial separation of the herbicide and the crop plant. Eg. Under canopy spray ensures contact of herbicide only with the weeds.
- The different germination or seed depths of weed and crop plant can lead to a soil applied (superficially) herbicide being absorbed only by weeds in the upper soil surface.

- Fixing of an agent can be caused by its low water solubility and interaction of herbicide with soil can be a factor. Depending on soil structure and texture and organic matter the herbicide may be adsorbed in the soil to a high or low degree.
- Difference in the rate of transport of soil-applied herbicides, inhibitory photosynthesis in weeds and crops lead to different concentrations in shoots and leaves and thus contribute to selectivity.
- Morphological difference among the various plants also influences the selectivity of a herbicide to a great extent.
- Biochemical causes like species-specific detoxification mechanisms are responsible for tolerance of plants to herbicides. Inactivation within the plant to non-phytotoxic derivatives or activation in a similar way determine the selectivity.

Types of Herbicides

Herbicides are broadly classified into selective and non-selective. Selective herbicides (eg. Butachlor, Atrazine) kill weeds alone selectively without affecting the crop. Non-selective herbicides (eg. Paraquat) kill both crop as well as weeds without any differentiation. These herbicides are applied normally in noncropped fields and in wide spaced crops like cotton, sugarcane, banana etc.

Herbicides are manufactured in four different forms viz

1. Soluble salt
2. Wettable powder
3. Emulation and concentrate
4. Granules

    - Soluble salts type of herbicides are completely soluble in water eg. 2,4-D sodium salt.
    - Wettable powder: Wetting agent is added and it enables the herbicides to form stable suspension in water. Eg. Atrazine, Diuron.
    - Emulsion concentrates: These can be mixed with water and sprayed or they can be mixed with sand and applied. Eg. Fluchloralin, Pendimethalin. They are in liquid form.

- Granular herbicides can be directly applied to the soil. They do not require sprayer for application. They do not come in contact with the crop foliage and there is no drift in the field.

Time of Application

Herbicides are applied during three different growth stages.

1. Pre-sowing or Pre-planting Herbicides

These herbicides are used to control the growth of weeds in the field before sowing of the crop plants. Eg. Paraquat is sprayed to kill weeds before taking up rice planting especially under zero or minimum tillage. Fluchloralin is sprayed and incorporated in soil in groundnut under rainfed conditions. Similarly in onion, bhendi and tomato, Fluchloralin is sprayed as a pre-planting herbicide. After spraying the herbicides the entire field is irrigated and seedlings are planted.

2. Pre-Emergence Herbicides

Most of the herbicides used are of pre-emergent type. These herbicides are sprayed during third to fifth day after sowing before the weed seeds germinate. Eg. Atrazine is sprayed third day after sowing in sorghum or maize.

3. Post-Emergence Herbicides

These herbicides are used to control grown up weeds. Eg. 2,4-D. Sodium salt is sprayed after 15 days after sowing in sorghum to control weeds.

Mode of action of herbicides

Contact and systemic.

Contact

- Weakening and disorganization of cellular membrane.
- Acute toxicity
- They undergo little/no translocation

Examples

PCP, Dinoseb, Paraquat, Cacodylic acid.

Systemic

- Affect one or more physiological/metabalical processes.
- Cause chronic toxicity.
- Translocated within the plant by transpiration stream (xylem) or by assimilation stream (Phloem).
- Penetrate cell membrane and accumulate at the site of action in toxic amounts without disturbing living system while in transit.
- They may be adsorbed/translocated in another form and their toxic effect induced in some other molecular form.

Examples

TCA, Dalapan, 2,4-D, 2,3,6, TBA Diuron, Linuron, Flumeturon, Siduron, Atrazine, Propazine, Prometone, Simazine.

Physiological process affected

1. Cell division
2. Tissue development
3. Chlorophyll and plastid formation

Cell division

- Herbicides are called as mitotic poisons.
- They may affect the processes at a particular stage such as between metaphase and anaphase and prevent cell wall formation resulting in multinucleate cells.

Tissue development

- Hormone type herbicides.
- Cause abnormal development and growth of cellular tissues such as xylem and phloem which interfere with the normal distribution of metabolites.

Chlorophyll and Plastic Formation

- Prevent Chlorophyll formation

Metabolic Processes

- Photosynthesis
- Respiration

- Nitrogen metabolism
- Enzyme activity

Photosynthesis

- By blocking the normal pathway of electron transfer-light reaction
- Some like paraquat and diquat act as electron acceptors and form free radials. Such free radials are oxidized to H_2O_2 which when accumulated destroy the cells.

Respiration

- by inhibiting electron transfer
- by inhibiting energy transfer
- by acting as uncoupling agents

Nitrogen Metabolism

- by inhibiting RNase and DNase
- by increasing the amino acid and protein contents

Enzyme activity

Alter the enzyme activity

- by changing the molecular structure
- by competing with substrate for active sites
- by forming complexes with the enzymes or substrate
- by competing with co-factor/changing the state of co-factor.

Herbicides-Organic Herbicides-Classification

1. Aliphatic-Carboxylic
2. Aromatic-Carboxylic
3. Phenol herbicides (dinitrophenol)
4. 2,6 dinitro aniline
5. Benzonitrile
6. 4,6 diamino-s-triazine
7. Phenyl urea
8. Phenyl Carbamate
9. Thiocarbamate

10. Organic arsenical
11. Uracil
12. Diphenyl Ether
13. Representative individual herbicides not yet members of herbicide family.

Eg.

Amibole	Glyphosate
Bentazon	Methyl Bromide
Desmedipham	Nor Flurazon
Dequat	Phenmedipham
Paraquat	Picloram

These besides inorganic herbicides like ferric sulphate, sulphuric acid, copper sulphate are also available.

Members of different groups

I. Aliphatic-Carboxylic

- TCA : Trichloro acetic acid-systemic
- Dalapan (Sodium 2,2 dichloropropionate)-Systemic
- Na TCA (Sodium Trichloroacetate)
- Kenapon

II. Aromatic-Carboxylic

Phenoxy-carboxylic herbicides	(i) 2,4-D
Silvex	(i) Dichloropop MCPA (ii) 2,4, DB Mecoprop (iii) 2,4,5, T MCPB
Phenyl-acetic acid herbicides	(i) Fenac
Benzoic acid herbicides	(i) Chloramben (ii) 2,3,6, TBA-Systemic
Phthalic acid herbicides	(i) DCPA (ii) Naptalam (iii) Endothall

III. Phenol Herbicides

Phenol

PCP (Pentachlorophenol) NaPCP

DNOC (4,6 Dinitro-ortho-cresol) Na DNNOC
Dinosam (DNAP) Amyl dinitrophenolate
Dinoseb (DNBP) Butyl Dinitro phenolate

Methyl	Benefin	Isopropalin	Sulfanilines	Nitralin
Anilines	Dibutalin	Pendimethalin		Orysalin
	Dinitramine	Profluralin		
	Fluchloralin	Trifluralin		

IV. Benzonitrile Herbivides

Dichlobenil
Bromoxynil

V. Diamino-S-Triazine Herbicides

Atrazine	Procyazine	Cyprazine
Cyanazine	Propazine	

Atrazine, Propazine, Simazine, Prometone are systemic.

VI. Phenyl Urea Herbiides

Fenuron	Linuron
Monuron	Flumeturon
Diuron, Systemic Siduron	Norea etc.

VII. Phenyl Carbamate Herbicides

1. Phenyl carbamates

Carbetamide	Karbutilate
Barban	Terbutol etc
Asulam	

2. Thiocarbamates

Thiobencarb or Benthiocarb	Cycloate
FPTC	Ethiolate
Vernolate	Pebulate
Molinate	Triallate etc.

3. Dithiocarbamates
   CDEC
   Metham

VIII. Acid Amide-Herbicides

Alachlor
Butachlor
Delachlor
Pronamide
Prynachlor etc.

IX. Organic-Arsenical

AMA, Cacodylic acid
CMA, DSMA, MAMA, MSMA

X. Uracil-Herbicides

Bromocil
Isocil
terbacil

XI. Diphenyl Ether Herbicides

Bifenox
Fluorodifen
Nitrofen
Nitroflurofen
Oxyfluorafen

Herbicides-Properties and Use

1. Aromatic-Carboxylic

(a) 2,4-D : 2,4-Dichlorphenoxy acetic acid

2,4-D can be used for destruction of dicotyledonous plants and it is well tolerated by many monocotyledonous crop plants.

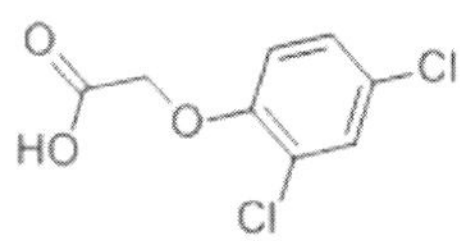

2,4-D is used as a selective herbicide particularly in cereals. The great advantages were cheap manufacture and relatively low mammalian toxicity. It is a systemic herbicide.

(b) 2,4,5, T : 2, 4, 5, Trichlorophenoxy acetic acid

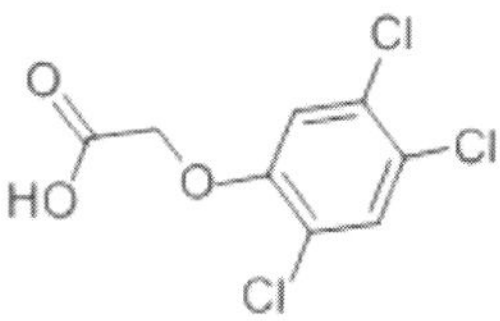

2,4,5, T has particularly high activity against woody plants and usually used in combination with other herbicides for control of trees, shrubs and control of otherwise intractable broad leaf weeds.

Structural analogues of IAA such as NAA or 2,4-D promote growth in lower doses but have a herbicidal effect at higher concentrations, 2,4,5, T is more persistent in soil than 2,4-D or MCPA (2-methyl-4p-chlorophenoxy acetic acid).

Mode of Action

- Interference with nucleic acid metabolism
- Disruption of translocation systems

2. Acid Amide Herbicide

a) Alachlor (Lasso, Monsanto)

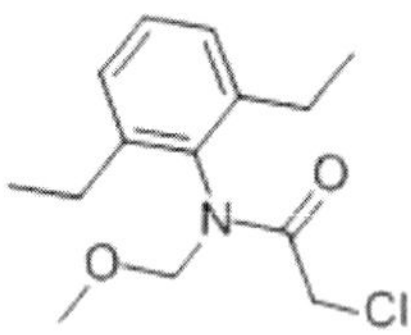

Pre-emergence against annual grass and some broad leaf weeds in corn, soybean, cotton, sugarcane, peanut, rape and brassicas.

LD_{50} : 1200 mg kg^{-1} rat

b) Butachlor (Machete, Monsanto)

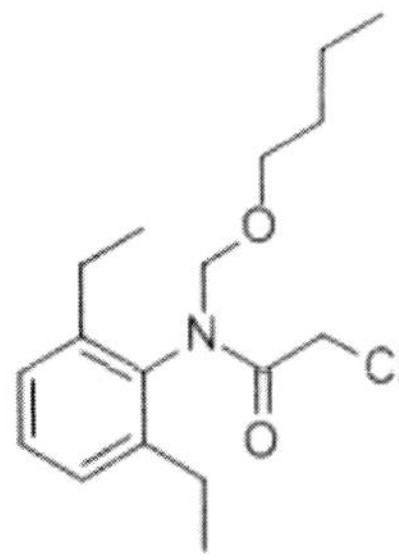

Pre-emergence against annual grass and some broad leaf weeds in rice, wheat, barley, sugarbeet, cotton, peanut and brassicas.

Alachlor and Butachlor are herbicidal chloroacetamides.

Mode of action : Inhibition of Photosynthnesis

Contact toxicity

Inhibition of cell division

3. Dinitro niline Herbicides

a) Fluchloraline (Basalin)

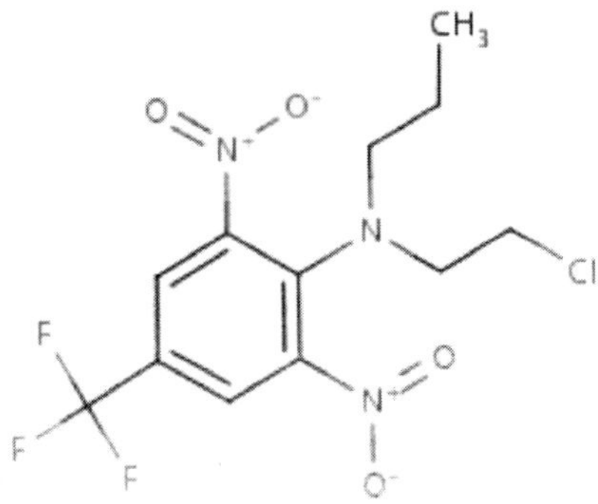

Against annual grasses and broad weeds in cotton and soybean; plant incorporated.

LD_{50} : 1150 mg kg^{-1} rat

b) Pendimethalin (Stomp)

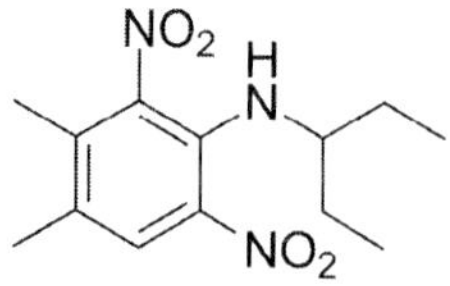

It is mainly used for the control grasses in cereals, rice, cotton, soybean, peanut and beans. Sowing; direct incorporation.

These are derivatives of dinitroanilines.

Mode of Action

The dinitroanilines are known to interfere with photosynthesis and respiration. They also act as mitotic poison. They inhibit root and shoot development by interfering with cell division.

4. Diphenyl ether Herbicide

Oxyfluorafen (Goal)

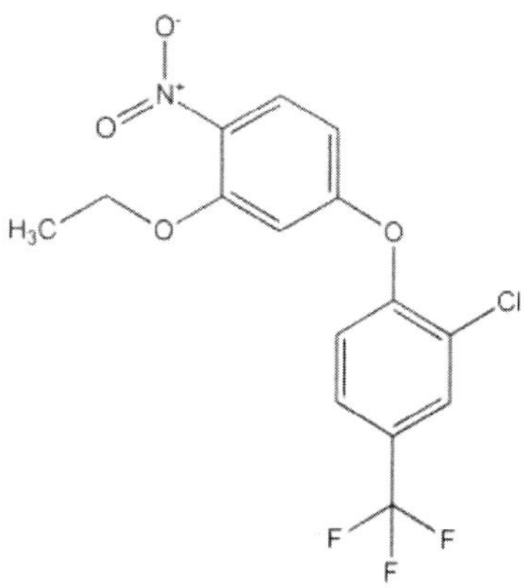

Used as pre and post emergence against annual broad leaf grass weeds in soybean, rice, cotton, peanut, tree fruit, nuts many tropical plantations ornamentals.

LD_{50} mg kg^{-1} rat

This is a derivative of diphenyl ether. The nitrodiphenyl ether herbicides are only.

Active in the presence of light and cause chlorosis of leaf tissue. They inhibit the Hill reaction in photosynthesis and photophos-phorylation.

5. Diamio-S-Triazine Herbicides Atrazine (Gesaprim)

Selective in corn, sorghum, sugarcane and pine apple.

Algic LD_{50}: 3080 mg kg^{-1} rat

Atrazine is effective under dry condition.

Atrazine is strongly adsorbed on various soil types remaining in the upper soil layers where they are taken up by germinating seeds. Deep rooted crops do not come into contact with the herbicide.

Many varieties of maize and sugarcane detoxify atrazine by hydrolysis in the plant tissues and hence these crops are resistant to the herbicidal action of atrazine. Therefore atrazine is very valuable for selective weed control in these crops.

Mode of Action

Inhibition of photosynthesis by interference with CO_2 fixation.

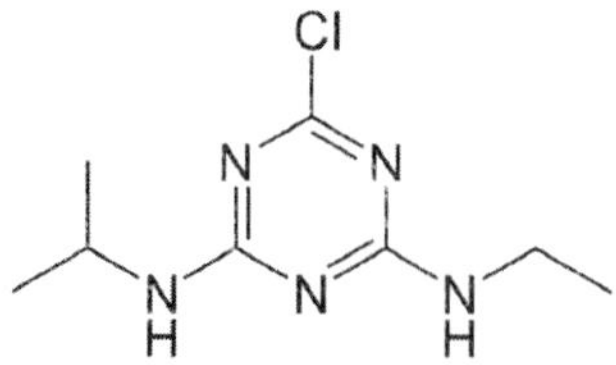

6. Individual herbicides

a) Paraquat (Gramoxone)

LD_{50}: 150 mg kg^{-1} rat

This is a bipyridylium salt derivative synthesized from Pyridine. Paraquat and Diquat are rapidly acting contact herbicides and desiccants that affect only aerial plant parts. In the soil they are strongly adsorbed and inactivated. Paraquat may be used against both broad leaf and grass weeds. The main use of paraquat is for total non-residual weed control.

Paraquat is also used for weed control in coffee, orchard fruit, and grape crops and for control of aquatic weeds (due to its low toxicity to fish).

Paraquat is active only in the presence of both light and oxygen.

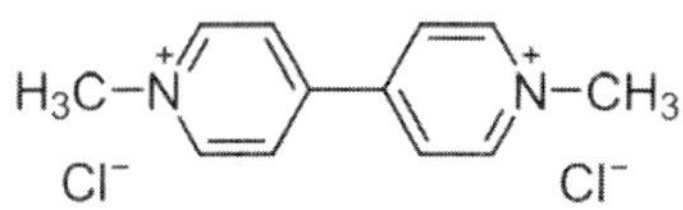

Mode of Action

Treatment with paraquat causes increased levels of hydrogen peroxide, which can readily diffuse through plant membranes.The hydrogen peroxide is the essential toxicant and kills the plant by lipid peroxidation-destroying cellular membranes.

b) Glyphosate (Round up)

The most important organophosphorus compound having herbicidal action is glyphosate. It was introduced in 1971 by Monsanto and is obtained by phosphorus trichloride and formaldehyde.

LD_{50} : 4320 mg kg^{-1} rat.

Glyphosate is a very potent non-selective post-emergence herbicide which kills mono and dicotyledonous annual and perennial weeds. It is rapidly translocated in the phloem from the foliage to the roots or rhizomes and hence effective against deep rooted perennial weeds. Eg. Cough grass and brambles at 0.7-5.6 kg/ha. Glyphosate is rapidly adsorbed on contact with soil and inactivated similar to paraquat. It is very effective against deep-rooted species of grasses, sedges and broadleaf weeds.

Mode of Action

The slow death of plants treated with glyphosate results chiefly from the reduction in protein synthesis arising from depletion in the supply of phenylalanine.

Glyphosate is a derivative of amino acid glycine. It was introduced by Monsanto as "Round up" in 1971. It is used as post-emergence and is rapidly absorbed by the leaves and translocated from vegetative parts to underground parts, rhizomes or stolons of perennial grass and broad leaved weeds.

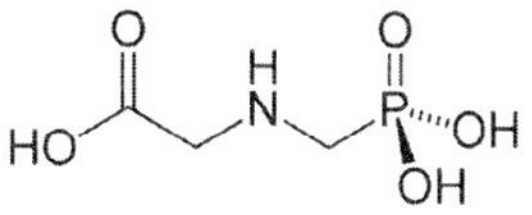

It can also be used in non-agricultural areas for bush control in forestry.

c) Anilofos

Anilofos is a member of organophosphorus group of herbicide which can be used as a soil applied post-emergence herbicide.

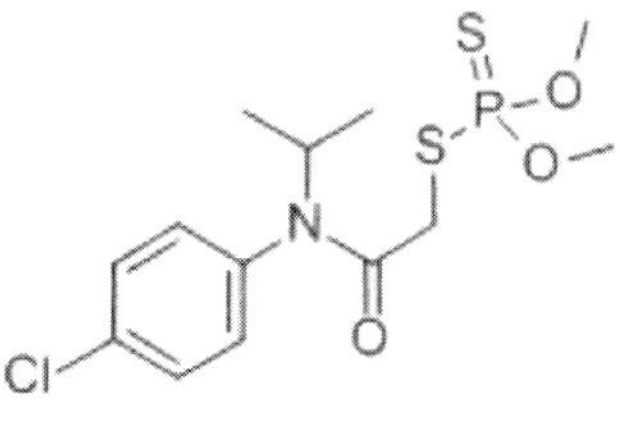

7. Phenylcarbamates-Thiocarbamate herbicides

Thiobencarb or Benthiocarb (Saturn, Bolero)

Used in rice, soybean, peanut, potato, beet, tomato.

LD_{50}: 1300 mg kg^{-1} rat

Thiobencarb

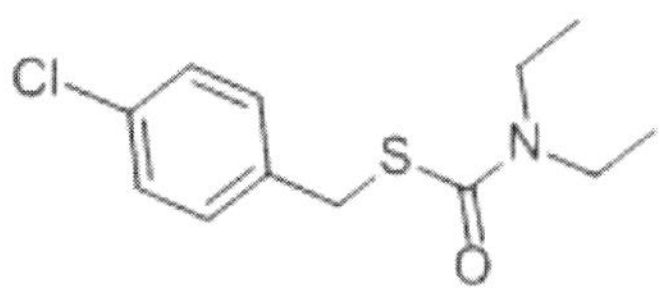

Soil-applied herbicide for pre-emergent control.

Mode of action

Not clearly determined. Inhibits growth in the meristematic region.

Thiocarbamates have a low aqueous solubility. They are most effective against grasses especially Nutgrass. Thiocarbamates are broad-spectrum herbicides.

Thiobencarb was introduced by Kumiai Chemicals as Saturn and by Chevron Chemicals as Bolero. It is an important herbicide in rice and barnyard grass. In addition it controls many other grass cyperaceous and broad leaved weeds.

❑❑❑

Chapter 32

Chemistry of Rodenticides

Under certain circumstances many vertebrates, even elephants, can rank as pests, but among mammals, the most serious are the rodents (Squirrels, rats and mice). The majority of poisons effective against vertebrates pests are also very toxic towards man and domestic animals. However they can be applied in such a way that they are not consumed by man or domestic animals.

Three rodent species are the major pests of agriculture, stored produce and present a potential danger to public health throughout the world. They are,

- Brown or Norway rat, *Rattus norvegicus*-major pest in temperature regions.
- Black Rat *Rattus rattus*-a dominant pest in Warmer regions.
- House Mouse-economically less important pest but lives in association with man in his dwellings to a greater extent.

Rodenticides

Rodenticides may be classified either as acute (single dose rapid acting) poisons or as chronic (multiple dose slow acting) poisons. Chronic poisons are taken at present only to included anticoagulants and have advantages in efficacy and safety but are slower to use and more expensive. Acute poisons have the advantage of rapid kill but may be too hazardous. Ideally a candidate rodenticide needs the three attributes of toxicity, acceptability and safety in use.

Chronic Poisons

The chronically acting or multiple dose rodenticides are inhibitors of blood platelet coagulation that reach their full effect after consumption of several sub-lethal doses.

As competitive antagonists of Vitamin K1, even small quantities of anticoagulant displace vitamin K1 from its role in the formation of prothrombin, which is necessary for blood coagulation. Prothrombimopenia is intensified by repeated doses and eventually the blood can no longer coagulate. Internal bleeding occurs in organs and tissues leading to an inconspicuous death. The rodent has no perception of the incipient poisoning and thus does not develop bait shyness (Antidote is vitamin K1).

The anticoagulants are coumarin and indanedione derivatives. The most important commercial products are

Warfarin
Coumachlor
Coumafuryl
Coumatetralyl
Warfarin (WARF 42)
Coumachlor (Tomorin or Ratilan)
Coumafuryl (Fumarin or Cumarax)
Copumatetralyl (Racumin).

Warfarin is prepared by condensation of 4-hydroxy coumarin with benzalacetone. Warfarin and other similar compounds function as anticoagulants so that the rats eventually die from internal haemorrhage after coming into the open in search of water. A dose of 1 mg/kg for five days is lethal.

Normally these anticoagulants control rodents effectively within two to five weeks and are relatively safe.

The chloroderivative of Warfarin known as coumachlor has similar anticoagulant properties. Coumarin derivatives coumatetraly, Difenacoum, Brodifacoum and Bromodiolone function as chronic rodenticides similar to Warfarin but more active (LD_{50} : 1.8).

Broadifocoum is specially useful because it is highly toxic to rats and mice but less toxic to man and animals.

Bromadialone is a very palatable and powerful anticoagulant which can achieve the kill of rats by one day's feeding of bait containing 0.0005% of bromodiolone (LD_{50} : 1.1%).

The trifluoromethyl compound is more active than other members of this group.

The newer second generation anticoagulants like Brodifacoum and Bromadiolone are currently the most commercially important rodenticides.

Several substituted indanediones also interfere with blood coagulation eg. Pindone, Chlorophacimnone.

Acute Poisons

Acute or single dose rodenticides are fully effective when only consumed once. The rapid onset of poisoning symptoms causes bait shyness in rats.

Aluminium phosphide (ALP), phosphide (Zn_3P_2) as well as Barium carbonate are used as acute poison rodenticides.

Aluminium phosphide pellets evolve poisonous phosphine gas and it is used to control most rats in paddy fields and both aluminium phosphide and sodium cyanide are effective against moles by placement in their burrows.

Moist zinc phosphide emits an unpleasant smell which repels most animals but it is surprisingly well accepted in baits by the generally suspicious rat. Zinc phosphide also emits phosphine gas when it comes in contact with dilute acid (gastric juice). LD_{50} : 40-50.

The most effective method of control of rodents now is the introduction of a powdered mixture of sodium cyanide, magnesium carbonate, and anhydrous magnesium sulphate (Cymag) and the powder, liberates lethal hydrogen cyanide gas in contact with the moist soil.

Other newer compounds

Norbromide, scilliroside, crimidine, chloralone, phosacetin, pyriminyl, Thiosemicarbazide, Chlorosilatran, Diphenylamine, Flupropadine.

Norbromide, another acute poison is a selective toxicant for the brown rat. Scilliroside is highly toxic again to brown rat. Crimidine and chloralone are other acte poisons. The narcotic chloralone used for control of bird pests can also be used against mice. Vacer, an alternative to the anticoagulants kills rats within a few hours through disruption of vitamin B metabolism.

❑❑❑

Chapter 33

Chemistry of Nematicides

Chemical products used for control of nematodes are termed nematicides. Nematodes appear in various shapes and forms on practically all crops.

The nematodes are unsegmented eelworms. Nematodes generally live in the soil, feeding on plant roots and some species invade stem and leaf. The most economically damaging is the root eelworm of potato which can survive as cysts for several years. The stem and leaf nematodes are second in importance to the root parasites and their control is easier since they appear only on aerial plant parts. They are usually killed by insecticides especially systemic insecticides.

In order to achieve effective control of nematodes the chemical should have the following properties.

- A nematicide must be able to distribute itself well in the soil.
- A nematicide should persist in active form in the soil for 1-4 months i.e., it should be long lasting in nematicidal action.

The nematicides can be grouped into two classes.

I. Fumigants and
II. Water soluble Agents.

Both classes have an unspecific pesticidal action, soil insects, soil-borne fungi and weed seeds being killed as well. They have very high phytotoxicity and hence can be used after the harvest of the crop.

Most important fumigants are as follows.

- Methyl Bromide; (Meth-O-Cas)
- Ethylene Dibromide (Bromafume, Nephis, Dowfume, Soil-Brom)
- 1,3-Dichloropropene (Telone; Vidden)

- A mixture of 1,3, Dichlorophropene and 1,2, Dichloropropene (DD)
- 1,2 Dibromo-3-chloropropane (Nemagon; Fumazone)
- A mixture of 1,3, Dichloropropene and 1,2, Dichloropropene (DD).
- 1,2 Dibrome-3-chloropropane (Nemagon; Fumazone).
- A mixture of chloropicrin and methyl isocyanate (Di-Trapex CF)
- Metham-Sodium (Sodium-methylcarbamodithioate) (Vapam)

Methyl Iso Thiocyanate (MIT)

Metham-Sodium or SMDC

Metham-sodium is applied in soil sterilization and is highly toxic to germinating weed seeds and nematodes; it is also used as a water soluble SMDC, which gradually evolves the active fumigant MIT (Methyl Isothiocyanate).

CH_3-N = C = S

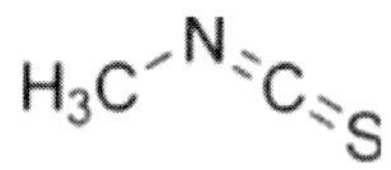

Methyl Iso Thiocyanate (MIT)

Water Soluble Agents

Chemical compounds that spread via water-capillary system of the soil are taken up by roots (systemic action) and poison the nematodes. In contrast to fumigants the toxic action of these compounds is specific. Both the phosphate and carbamate types of nematicides are inhibitors of cholinesterase.

The most important organophosphate and carbamate nematicides are Dichlofenthion, Fensulfothion, Fenamiphos, Diamidafos, Carbofuran, Aldicarb, Phorate and Disulfoton.

Phorate

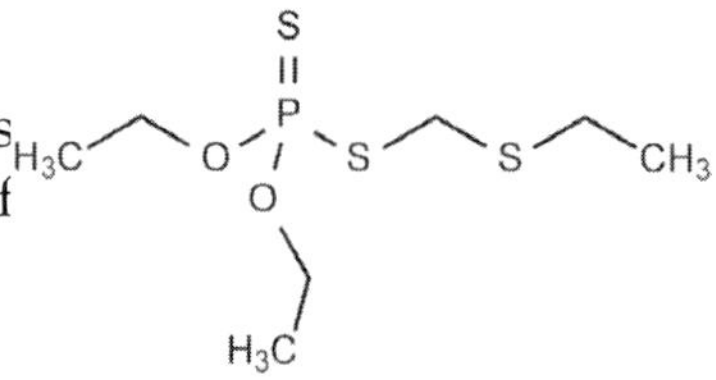

Phorate is systemic. Disulfoton is formulated as granules for control of aphids and other like nematodes.

Fensulphothion (Terracur-P)

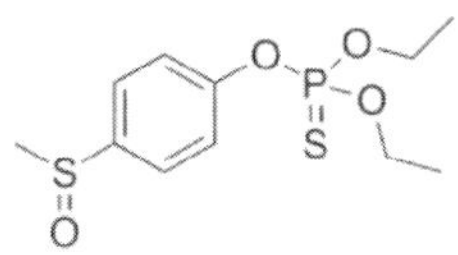

A nematicide and insecticide with a persistence of 4-6 months.

LD_{50} : 2-10 mg kg^{-1} rat.

Fenamiphos (Nemacur)

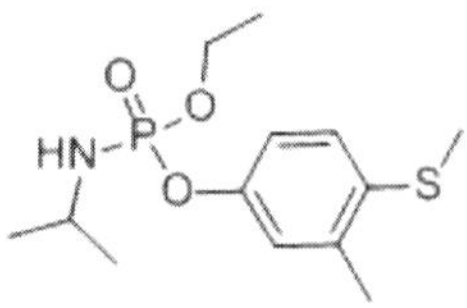

A particularly valuable nematicide for control of free-living, cyst forming and root-gall nematodes in sugarbeet and potatoes.

LD_{50} : 15-19 mg kg^{-1} rat.

Aldicarb (Temik)

Systemic insecticides and nematicide.

LD_{50} : 0.93 mg kg^{-1} rat.

Carbofuran (Furadan)

Highly active insecticide, Acaricide and Nematicide.

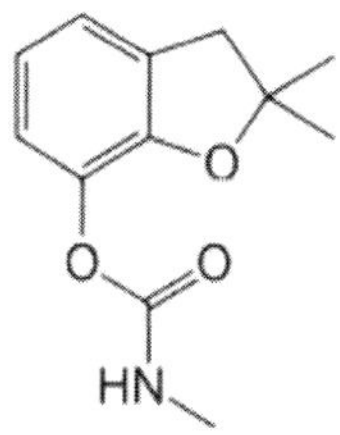

Chapter 34

Pesticides Control Order

The hazard of pesticides can be prevented by regulation of marketing, evaluating their properties, performance and hazards of the products through legislation to provide maximum protection to the society from the adverse effects.

Need for Legislation

- Prevention of misuse
- Protection of workers in preparing, transporting and applying.
- Prevention of contamination of food stuffs.
- Maintenance of acceptable residues.
- Protection of wild life and prevention of pollution.

Legislation in India

A comprehensive bill was passed in the Parliament in 1968 known as Insecticides Act 1968 and the provisions of the Act were brought into force in 1971. The act was amended in 1972 and in 1977.

The Insecticides Act 1968

An act to regulate the import, manufacture, sale, transport, distribution and use of insecticides with a view to prevent risk to human beings or animals and for matters connected therewith.

Salient Feature

- Compulsory registration of the product at the central level and Licences for manufacture and formulation at state level.
- Central insecticides Board for Co-ordination.
- Registration Committee for Registration of all insecticides.
- Analysis and inspection by Central/State Government.
- Establishment of Central Laboratory.

- Regulation of import and manufacture.
- Framing rules, Forms and Fees.

The Central Insecticides Board

Advises on matters relating to

- The risk involved to human beings and animals and safety measures for protection.
- The manufacture, sale, storage, transport and distribution of insecticides.

The Registration Committee

The Central government constitutes a Registration Committee.

- To register insecticides after scrutinizing their formulae, efficacy and safety.
- To perform other functions as per the Act.

Licensing

The State Government appoints Licensing Officers for the purposes of the Act and defines areas of jurisdiction.

Any person desiring to manufacture, sell, stock, exhibit or distribute any insecticide may make an application to the licensing officer for the grant of a license.

Central Insecticides Laboratory

Central Government establishes a central insecticides laboratory to carry out the functions of analysis of pesticides. Insecticides analysts are appointed by central/state governments.

Insecticide Inspectors

Are appointed by Central/State government for specific areas. An insecticide inspector is empowered.

- To enter and search premises for any offence under this act.
- To stop distribution, sale or use of insecticide in case of violation of this act.
- To take samples of insecticides and send them for analysis to the insecticides analyst.

The Insecticides Rules 1971

There are 9 characters in the insecticides rules, 1971. These rules cover the aspects of definition of terms, functions of Board, Registration Committee and Laboratory, Grant of Licenses, for manufacture and sale, packing and labeling.

These rules also cover the aspects of analysis, qualifications, duties, powers of insecticide analyst and insecticide inspector. These chapters of rules also deal with the transport, storage and provisions regarding protective clothing, equipment and other facilities for workers during manufacture etc., of insecticides.

Shelf-Life

The shelf-life of technical grade pesticides or their formulations have not been specified in the standards. The stability of the material is determined on the basis for the declaration of the expiry date by the manufacturer, which is obligatory under the insecticides act 1968. Shelf life or stability of any pesticide is an important factor which can be expressed qualitatively and quantitatively. Pesticides manufactured with maximum care and under the most ideal conditions cannot remain stable over a prolonged period because these are thermodynamically unstable. The central insecticide board after conducting trials on a pesticide under various conditions of temperature (climatic zones), transportation and storage and analyzing the date, approves a shelf life for the product. The Indian standards specify that storage stability tests (heat stability in the case of emulsifiable concentrates and accelerated storage in case of wettable powders) need not be carried out in case of the formulations which have completed half of the shelf life as per the manufacturing declarations.

List of Pesticides Banned / Restricted in India

S.No.	Banned	Restricted
1.	Aldrin	1. AluminiumPhosphide
2.	Chlordane	2. BHC
3.	Dibromochloropropane	3. Captafol
4.	Heptachlor	4. Chlorobenzilate
5.	Nicotine sulphate	5. DDT
6.	Nitrofen	6. Dieldrin

7.	Paraquat dimethyl sulphate	7.	Lindane
8.	Pentachloro Nitrobenzene	8.	Methyl parathion
9.	Pentachloro phenyl (PCB)	9.	EDB
10.	Phenyl Mercury Acetate	10.	Sodium cyanide
11.	Tetradifon		
12.	Toxaphene		
13.	Calcium cyanide		
14.	Copper Acetoarsenite		
15.	Ethyl Mercury chloride		
16.	Menazon		
17.	Sodium methanoarsenate		

❑❑❑

Chapter 35

Compatibility of Pesticides with Other Agrochemicals

Many of the pesticides cannot control all types of pests and diseases and hence it becomes necessary to use number of them together to control various pests and diseases. There is also considerable saving of labour, time and water especially in areas of water scarcity, if these could be mixed and applied in one single spray. This necessitates studies on the compatibility of pesticides with fertilizers and other chemicals. Pesticides, themselves, being chemicals, they tend to react among themselves as well as other agrochemicals. This is result in

- Losing its pesticidal activity,
- Becoming phytotoxic,
- Adversely affecting the spraying equipments like clogging the nozzle or corroding the sprayer.

Many pesticide firms carryout compatibility studies with the new candidate chemicals. A compatibility chart has been prepared by the Farm Chemicals Information, USA as a rough guide for mixing for pesticides.

Studies have indicated that

- Combinations of captan + dieldrin, lindane
- Furadan + Ziram
- Thiram, Malathion, Metasystox, Dithane M-45 are compatible and synergistic.

Combination of fertilizers

- Urea + thiram, phosphamidon, dithane-M45, thiometon, Endosulfan.
- Dechlorovos, parathion, carbaryl

- Neutral or acidic fertilizers with captan
- Superphosphate + Lindane
- Ammonium phosphate + rogor
- Ammonium sulphate + FYM + phorate are compatible. But
- Urea + Lime sulphur, Karnathane, Sevin, Morocide are not compatible.
- NP fertilizers + Rogor are not compatible since this combination is phytotoxic.
- Compatibility, however, also depends on mixing containers, mode of mixing, emulsifier, wetting agents and other additives used.

Compatibility is assessed by

- Germination percentage,
- Yield of crops,
- Phytotoxicity,
- Per cent to pests or diseases controlled.

❑❑❑

Chapter 36

Impact of Pesticides in Environment

Reactions of pesticides in soils

Whether pesticides are applied to plant foliage, to the soil surface, or are incorporated into the soil, a high proportion of the chemicals eventually moves into the soil. These chemicals then move in one or more of six major directions.

- They may vaporize into the atmosphere without chemical change.
- They may be adsorbed by humus and clay particles.
- They may move downward through the soil in liquid or solution form and be lost from the soil by leaching.
- They may undergo chemical reactions within or on the surface of the soil.
- They may broken down by soil microorganisms.
- They may be adsorbed by plants and detoxified within the plants.

The specific fate of these chemicals will be determined at least in part by their chemicals structures, which as shown in figure 18.2 are quite variable. Various processes affecting organic chemicals in the soil are represented in the following figure :

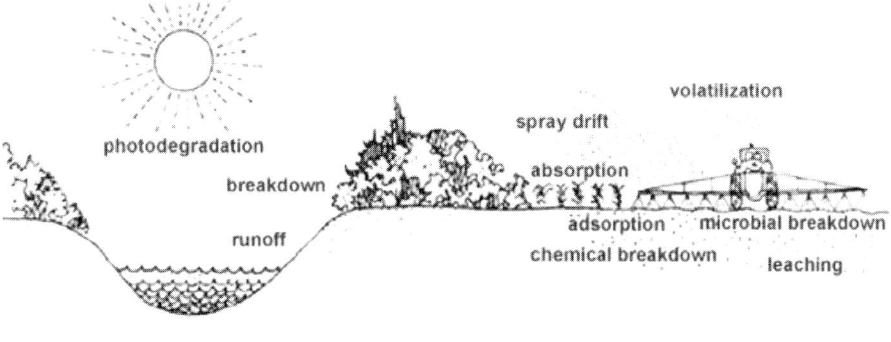

Volatility

Pesticides vary greatly in their volatility and subsequent susceptibility to atmospheric loss. Some soil fumigants, such as methyl bromide, are selected because of their very high vapour pressure, which permits them to penetrate soil pores to contact the target organism. This same characteristics encourages rapid loss to the atmosphere after treatment unless the soil is covered or sealed. A few herbicides (eg., trifluralin) and fungicides (e.g., PCNB) are sufficiently volatile to make vapourization a primary means of their loss from soil.

Earlier assumptions that disappearance of pesticides from soils was evidence of their breakdown are now known to be questionable. That some chemicals as lost to the atmosphere only to return to the soil or surface waters in rain is now known.

Absorption

The absorption of pesticides by soil is determined largely by the characteristics of the pesticides and of the soil to which they are add. The presence of certain functional groups, such as –OH, NH_2, -NHR, -$CONH_2$, - COOR and = - + NR_3, in the chemical structure, encourages adsorption, especially on the soil humus. Hydrogen bonding and protonation (adding H^+ to a group such as an –NH_2 (amino) group) probably promotes some of the adsorption. In general, the larger the size of the pesticide molecule, everything else being equal, the greater is its adsorption.

Some pesticides with positively charged groups, such as the herbicides diquat and paraquat, are also adsorbed by silicate clays. Adsorption by clays of some pesticides tends to be pH dependent, maximum adsorption occurring at low pH levels where H^+ bonding occurs. Addition of H^+ ion to functional groups (e.g., -NH_2) yields a positive charge on the herbicide, resulting in the attraction by the negatively charged soil colloids.

Leaching

The tendency of pesticides to leach from soils is closely related to their solubility and their potential for adsorption. Strongly adsorbed molecules are not likely to move down the profile. Likewise, conditions that encourage adsorption will discourage leaching. Leaching is apt to be favoured by water movement, taking place most readily in permeable sandy soils that are low in clay and organic matter.

In general, herbicides seem to be more mobile than either fungicides or insecticides and their extensive use has resulted in groundwater pollution in some areas.

Chemical Reactions

Some pesticides undergo chemical modification independent of soil organisms. For example, diquat and the triazines are subject to slow photodecomposition activated, the triazine herbicides (e.g., atrazine) an organophosphate insecticides (e.g., malathion) are subject to hydrolysis and subsequent degradation in the soil. While the complexities of molecular structure of the pesticides suggest different mechanisms of breakdown, it is important to realize that degradation independent of soil organisms does in fact occur.

Microbial Metabolism

Biochemical degradation by soil organisms is the single most important method by which pesticides are removed from soils. Certain polar groups such as -OH, COO- and NH_2 on the pesticide molecules provide points of attack for the organisms.

Chlorinated hydrocarbons such as Lindane and Endosulfan are very slowly broken down in most soils. In contrast, the organophosphate insecticides such as parathion are degraded, quite rapidly, apparently by a variety of organisms. Likewise, the most widely used herbicides, such as 2,4-D, the phenylureas, the aliphatic acids, and the carbamates, are readily attacked by a host of organisms. Exceptions are the triazines (e.g., atrazine), which as degraded slowly, primarily by chemical action. Most organic fungicides are also subject to microbial decomposition, although the rate of breakdown of some is slow, causing trouble some residue problems.

Plant Absorption

Pesticides are commonly absorbed by higher plants a necessary process for the effectiveness of most herbicides and of some insecticides. The absorbed chemicals may remain intact inside the plant or they may be degraded. Some degradation products are quite harmless to humans and other creatures, but others are toxic, in some cases even more toxic than the original pesticide. Pesticide residues, especially those found in the edible portion of plants (fruits, nuts, etc.), are of critical concern to humans. Such residue levels are strictly regulated by law to assure human safety.

Persistence in Soils

The persistence of pesticides in soils is a summation of all the reactions, movements and degradations affecting these chemicals/ marked differences in persistence are the rule. For example, Organophosphate insecticides may last only a few days in soils. The most widely used herbicide, 2,4-D, persists in soils for only 2-4 weeks, whereas chlorinated hydrocarbons may persist far from 3 to 15 years or longer. The persistence times of other herbicides, fungicides and insecticides generally are between the extremes cited. The majority of pesticides degrade rapidly enough to prevent buildup in soils. Those that resist degradation have potential for environmental damage.

Continued use of the same pesticide on the same land can increase the rate of microbial breakdown of the chemical. This is an advantage in relation to environmental quality, but the breakdown may become sufficiently rapid to reduce the herbicide's effectiveness.

Reducing Soil Pesticides Levels

Among the practices suggested to reduce pesticide levels in soils is the addition of easily decomposed organic matter. The growth of high-nitrogen cover crops or the additions of large quantities of animal manure's also should be helpful. Apparently degradation of even the most resistant pesticides is encouraged by conditions favouring overall microbial action. Other practices suggested to reduce pesticide and leaching the soil. Unfortunately, some of these procedures merely transfer the chemical from the soil to some other part of the environment, a process of dubious value.

This brief view, of the behaviour of pesticides in soils reemphasizes the complexity of the changes that take place when new and exotic chemicals are added to our environment. Clearly the ecological effect of new chemicals must be evaluated as thoroughly as possible before their extensive use is permitted.

❑❑❑